JN064571

IOWN

Jun Sawada President & CEO, NTT Corporation [Supervising Editor]
Motoyuki Ii & Katsuhiko Kawazoe

Innovative Optical and Wireless Network

IOWN

Beyond the Internet

NTT Publishing Co.

NTT Publishing Co., Ltd.

NTT Publishing Co., Ltd. is an associated company of
Nippon Telegraph and Telephone Corporation.

© 2020 Nippon Telegraph and Telephone Corporation

Published by NTT Publishing Co., Ltd.
3-4-1 Shibaura, Minato-ku, Tokyo, 108-0023, JAPAN
www.nttpub.co.jp

Designed by Yukimasa MATSUDA + Shouji SUGIMOTO

ISBN: 978-4-7571-0395-5

Preface
Moving into the IOWN Era

Jun Sawada
President and Chief Executive Officer, Member of the Board,
NTT Corporation

IOWN (/aiən/: Innovative Optical and Wireless Network) will serve as the infrastructure for realizing the future world that we at NTT are striving for, together with our partners. This concept is currently under review with the aim of finalizing specifications in 2024 and achieving early commercialization. We decided to profile IOWN in this book because we wanted to showcase the world our IOWN initiative is aiming for, and the technologies for making it a reality.

Looking back on history, they say that 17th century Japan was under a policy of seclusion, but in fact we were disseminating technology and culture to the world. Cloths resembling Japanese Kimonos appear and other Japanese products appear in the paintings of the Dutch artist Vermeer, and this shows how they had penetrated into daily life in Holland through trade. Silver obtained through a monopoly on trade with Japan is said to have contributed to Holland's economic prosperity at that time. Now, in the 21st century, we have conceived the IOWN initiative in the hope of once again disseminating Japanese innovation to the world.

The business environment surrounding Japan is currently undergoing huge transformations. As globalization accelerates with the explosive spread of the Internet, divisions between countries have surfaced due to the emergence of protectionism—as exemplified by trends like the US-China trade war and Brexit. We are deluged with information in the form of big data because everything is connected with the Internet, and fragmentation is intensifying between individuals due to factors like information filtering using AI. In response to these conflicts—global versus local, and centralization versus decentralization—we believe modern society needs to realize a paraconsistent world which accepts contradictions and connects both sides of each conflict. In this way, we can build a community where people mutually recognize diverse values.

To realize this sort of society, we will need new innovations that go beyond previous concepts. Achieving new innovations naturally requires progress from a technical standpoint, but new social perspectives are also needed, reflecting input from social science and the humanities (e.g., ethics, morality, law). IOWN we are working to realize is a new infrastructure for the information society. It will comprise an "All-Photonics Network" incorporating photonics-based technology into every aspect, "Digital Twin Computing" for tasks like predicting the future by combining the real world and the digital world, and a "Cognitive Foundation" for connecting everything together and exerting control. Realizing this infrastructure will require consideration from a wide range of perspectives, such as natural and applied science, social science, and the humanities, and we hope to discuss the relevant issues extensively with people from industry, both inside and outside Japan, and from academia in a diverse range of research and technology fields.

I hope those who pick up this book will understand our dreams and objectives, and work with us to create the next phase of the future.

Contents

Part 1

The World of IOWN

Chapter 1

What is the IOWN Initiative?

Katsuhiko Kawazoe

Senior Vice President, Head of Research and Development Planning, Member of the Board, NTT Corporation

1. Background of the IOWN initiative

ICT is accelerating change in daily life

People's daily lives have constantly changed throughout recorded history. Social structures changed from hunter-gatherer society to agricultural society, and then to industrial society. As this occurred, the human population repeatedly went through stages of growth and equilibrium, and there was always the pressing issue of limits on consumable energy, and conflict between people seeking a more affluent life. Throughout history, people have continually made progress toward achieving their desires, while producing new innovations.

The industrial revolution which began in the second half of the 18th century was a major turning point, and since then the speed of social change has accelerated. In one classification, the start of industrialization due to the invention of the steam engine is called the first industrial revolution; the start of mass production using electric power is called the second industrial revolution; the start of technical innovation based on information and communications is called the third industrial revolution; and today's era—exploiting big data, IoT (Internet of Things), AI, and robots—is called the 4th industrial revolution. Various frameworks for the new IoT society have been launched in different countries, such as Society 5.0 in Japan and Industry 4.0 in Germany.

In recent years, the Internet and the smartphone have been particularly important innovations. They're an indispensable part of our lives today, and in little more than a decade, have greatly changed the nature of society. One of the greatest achievements of the Internet is that it dramatically lowered transaction costs for information and goods, and the business environment has changed dramatically due to the development of various transaction platforms on the Internet. The smartphone makes the benefits of services offered via applications for such platforms available to people all over the world.

The informatization of society will likely continue to accelerate in the future, and bring about tremendous changes to our lives due to adoption of various forms of AI and IoT technology in the settings of daily life, e.g., new financial services (fintech) employing ICT (information and communications technology), and self-driving vehicles using AI.

Lifestyles anticipating these changes are already emerging among young people today. For example, there is a developing culture of excitement about esports, with fans admiring, respecting, and supporting players. The occupation of "professional gamer" is increasingly recognized, with world-level competitions and other events. A world already exists, linked with economic activity, where people communicate with each other in virtual spaces connected to the real world, work hard in their daily training, form communities, and seek self-realization in virtual space.

Values changing with society

These technology-driven social changes have also started to change people's values. For example, values relating to goods have changed, and a shift from the previous culture of ownership to a culture of use is underway, as seen in trends like sharing of automobiles and homes. Open science is being advocated as a forward-thinking approach to research, and there is a new and growing trend of sharing technologies and knowledge.

On the other hand, there has been progress in the "personalization of desire," with people and companies moving away from previous shared values, and looking for things that suit their inner selves. No longer satisfied with conventional mass production and mass consumption, an increasing number of people are choosing products that reflect their lifestyle, saving up and purchasing items they like, even if they are expensive, and paying a premium for added value that suits their tastes [1]. There are also trends like seeking out items that are old and have character, even if they are inconvenient; and ethical consumption (selecting and purchasing items that comport with ethics and morality) where people select things that are a little more expensive but good for the environment or contribute to society. In other words, people are not moving in one direction, seeking expansion and growth. They are starting to choose things they personally regard as good, based on diverse values.

There is a growing trend of emphasizing spiritual affluence rather than wealth of material goods in daily life[2]. The source of this, we might say, is a desire for the happiness and well-being of both individual people and society as a whole.

Due to these changes in values, we face questions like what goals we should adopt for the coming technological revolution, and what vision of society we should pursue as we move forward.

[1] Source: Changes in consumer values (seeking things that suit oneself): Nomura Research Institute "Changes in values and consumer behavior of the Japanese, as seen in a questionnaire survey of 10,000 consumers — Key points on the results of the 7th time-series survey" (In Japanese)

[2] Source: EY Institute Insight, Vol. 2, Autumn 2014 Report, "Changing values: From the era of sharing" (Cabinet Office, Public Opinion Survey Concerning Lifestyles of the Japanese People (June 2013 survey)) (In Japanese)

2. Social significance of the IOWN initiative

An information environment for promoting understanding of diverse people

Let's look, in a little more detail, at social changes from the standpoint of the information field. Already, 30 years have passed since dissemination of ICT, exemplified by the Internet, and there is a need to establish diverse lifestyles (values) on both the real and virtual side. At the same time, fragmentation is intensifying everywhere—at the personal, group, and national level. Various factors underlie this, including: a deluge of information so great it exceeds the cognitive information processing abilities of human beings, limitations of existing social institutions, and growth of information disparities. These factors are thought to foster a lack of understanding and apathy toward other people.

To remedy this problem, people's values will need to be nudged toward greater acceptance of diversity, and to support that, we will need to create a framework for society and the world as a whole, including things like institutional systems. That is, we need to create "information environment = place."

If in this "place" more information is distributed and processed in real-time—fairly and without discrimination—then that information will include diverse values, and facilitate sharing of the perspectives and experiences of others. And if we can thus promote social conduct based on understanding and empathy with others, then it may be possible to improve the quality of connections between people, and between people and society, and reaffirm the values of individuals as a result.

Striving to realize "information environment = place"

As one way to realize this framework of "information environment = place", we have proposed the IOWN (Innovative Optical and Wireless Network) initiative, and commenced efforts to achieve realization in 2030. In the IOWN initiative, the aim is to revolutionize previous information and communications systems, and realize a new information and communications infrastructure transcending the limitations of current ICT technology. We plan to achieve this through an "All-Photonics Network" employing optical technology; "Digital Twin Computing" implemented on top of that for performing analysis and feedback processing in real-time; and a "Cognitive Foundation" for allocating resources while balancing processing to achieve overall optimization, and distributing necessary information within the network. (Details are given below.)

Around 2030, many expect to see tremendous changes in society in a wide range of areas including self-driving vehicles, transportation, medicine, finance, and manufacturing. IOWN will overcome the issues of existing ICT technology, and spark a paradigm shift in society. These changes will not be limited to resolving hindrances in today's digital society. IOWN will create a world where people enjoy the benefits of sophisticated technology, in accordance with their different values and situations, without requiring any awareness of the underlying technologies.

Naturally, the new "information environment = place" approach for realizing this will not be a creation only of NTT, and it will not be limited only to Japan. We are widely recruiting participants to support this project, from both inside and outside Japan, and we intend to implement this initiative through open innovation.

We are also engaging with initiatives like the SDGs that aim to solve global social problems, as well as the Japanese national program Society 5.0. We will strive to develop people-friendly technology that can help in building systems to solve social problems.

Advanced technology sometimes has negative aspects that diminish the happiness or well-being of people. Therefore, we will also solicit broad participation from academia, involve experts from the humanities and social sciences as well as information science and engineering, and thoroughly discuss how to respond to the various changes brought on by new technologies.

3. 3 issues for the realization of the IOWN initiative

What will be needed to realize this new "place"? And what limits will have to be overcome? Here we will consider issues to be addressed in opening up this new world.

Issue 1: Dealing with diversity

In a richly diverse world, humans and all other things—including the natural world— coexist, and the individuality of each person shines out. What makes this possible is understanding of others. To deepen such understanding, it's very helpful to see information and sensibilities from the perspective of others different from oneself, and information through the eyes of others. This input may greatly change one's own values, and also lead to the realization that one is accepted by others.

To achieve this with technology, it is not enough to simply obtain more information by developing higher-accuracy, higher-sensitivity sensors. There is a need for information processing that overlaps with the sensibility and subjectivity of others. For that purpose, we will likely need to incorporate knowledge not only from science and technology, but also from the humanities and social sciences. In addition to gaining better knowledge of human beings themselves, there is a need to shift attention to the diversity of life and systems, without being limited by previous received wisdom.

We use the term "natural" for the agreeable condition that arises when human beings naturally enjoy the results of technology without stress. We also use the term "natural harmonic" for a world where people and the environment are in harmony. These will be our objectives.

Issue 2: Transcending the limits of the Internet

Realizing this sort of world will, of course, bring about a huge increase in the volume of information. It can be readily imagined that limits will emerge, in terms of both transmission and processing capacity, if we continue with existing information and communications systems.

The distributed amount of data has sharply increased over the last 30 years, during which smartphones and the Internet have been disseminated throughout society. Over about the last 20 years, from 2006, communications traffic volume per second is estimated to have increased by 190 times [3] (from 637 Gbps to 121 Tbps [4]) on the Internet in Japan. Over the 7 years from 2018, it is estimated that the total amount of data worldwide will increase from 33 ZB [5] to 175 ZB [6] (Fig. 1).

The Internet is a massive network, built by interconnecting multiple networks based on common communication procedures like IP (Internet Protocol). If we rely only on this network, we will face a variety of issues that cannot be

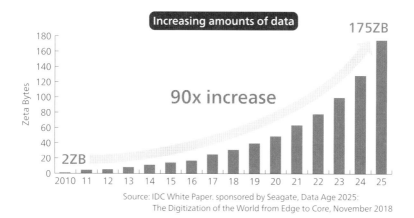

Source: IDC White Paper. sponsored by Seagate, Data Age 2025:
The Digitization of the World from Edge to Core, November 2018

Fig.1: Estimated increase in volume of data

solved by extending current technology, such as further increases in communications traffic volume, greater network complexity, and increased latency due to congestion.

[3] bps. b (bit) is a unit indicating a quantity of data. bps is a unit of transmission speed, i.e., sending 1 bit of data in 1 second. G (giga) indicates 10 to the 9th power, and T (Tera) indicates 10 to the 12th power.

[4] Ministry of Economy, Trade and Industry "Green IT Initiative" (2007) (In Japanese) https://home.jeita.or. jp/upload_file/20130502100819_DAxFoOrkXW.pdf

[5] Zetabyte. B (byte) is a unit indicating a quantity of data. 1 byte is 8 bits. Zeta indicates 10 to the 21st power.

[6] Data Age 2025, sponsored by Seagate with data from IDC Global DataSphere, Nov 2018

Issue 3: Overcoming the increase in electric power consumption

At present, there are two major causes of concern in terms of energy consumption: the explosive increase in the number of electronic devices connected to the Internet due to the growth of IoT, and the increasing load on the network **(Fig. 2)**.

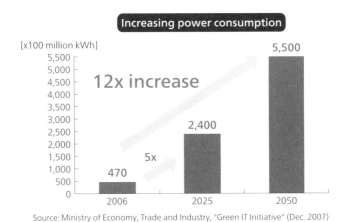

Source: Ministry of Economy, Trade and Industry, "Green IT Initiative" (Dec. 2007)

Fig.2: Estimated power consumption by IT equipment

Although progress is being made in reducing the energy consumption of each electronic device, energy-saving measures cannot keep up with the increase in quantity. The tremendous power consumption of the data centers which are essential for building information and communications systems and providing cloud services is also a world-wide problem. Even today, there are uncertainties about electric power supply, and conditions are such that we may have to forgo construction of new data centers for cloud services.

Another focus of great concern in recent years has been the breakdown of Moore's law. "Moore's law" was described in a 1965 article by Gordon Moore, one of the founders of Intel. Based on empirical observation of the semiconductor industry, Moore stated that "the number of transistors on an integrated circuit, per unit area, doubles every 18 to 24 months."

Transistors are components that form arithmetic and memory elements by amplifying and switching electrical signals, and they play a central role in information processing. An increase in the number of transistors signifies an improvement in performance, and a circuit with the same performance can be mass produced in a smaller chip area. This has helped hold down manufacturing costs.

However, the size of today's transistors has already been reduced to nm units (nanometers, 1 nanometer = one-billionth of a meter), and manufacturing is running up against physical limits. Furthermore, the increased degree of integration sometimes results in malfunction due to variation in the number of electrons flowing through an integrated circuit. Another effect is rising temperatures due to increased heat production. Limitations on operating frequency are also approaching.

We are beginning to see the limits on electronics technology supporting current ICT systems, and if we reach a point with no prospect of further progress, this may lead to stagnation not only of technology, but of society.

4. A paradigm shift to solve these 3 issues

As the key to solving these fundamental problems, we believe there is tremendous potential in shifting from "digital to natural" and from "electronics to photonics."

1. From digital to natural

A natural approach patterned after the natural world

In short, natural technology provides new methods of sensing and perception impossible with previous digital technology. Here we will present a number of approaches for such natural technology.

In the natural world, we can find analogies that help us appreciate diverse values. Ultraviolet rays are not visible to the human eye, but they are visible to insects and birds. Flowers have the characteristic of concentrating ultraviolet rays in their center, and bees can accurately find the center of a flower by using ultraviolet rays **(Fig. 3)**. For humans—who seek beauty in flowers—their value lies in beautiful color and texture, but for bees their value lies in accurately indicating the location of the nectar. In other words, different perceivers find value in different things, and the world appears different to them, even when looking at the same scene.

If—proceeding along these lines—we can create a world where we carefully extract and accumulate various types of information never acquired before because its value was not evident, transmit that information far and wide, and enable its use in information processing, then we may be able to provide people with various types of value through previously-unimaginable overlays of data.

In developing connected cars [7] and MaaS [8], value lies in the ability to accurately and instantaneously check "whether it is possible to safely travel this road." Here, there is no need to send data-intensive beautiful video to eyes that see with high definition. It is enough to determine whether the road is frozen or whether some obstacles have fallen into the road, i.e., whether the road is safe. In each situation, we want to understand the value sought by customers, and the

Appearance of flower to human vision   Science Photo Library, Aflo

To the vision of bees, the center part with   Science Photo Library, Aflo
honey boldly stands out

Fig.3

customers' values, and realize services which provide solutions autonomously in various forms.

Values naturally differ between cultures, nations, and individuals. Whenever there is conflict between nations or individuals, a difference in values is often a major contributing factor. If we can understand diverse values, and grasp them at a level deeper than competing sides, then a natural form of conflict resolution may be brought about.

Of course, ethical dilemmas regarding values, such as the trolley problem [9], inevitably arise here. Unfamiliar problems are sure to crop up at every turn— problems inconceivable under the conventional, unitary, mechanistic view of information. For this reason, knowledge from the humanities and social sciences will be essential for building an IOWN world consonant with our pluralistic values.

[7] Automobiles with functionality as ICT terminals. Various types of data, such as the state of the vehicle and the situation of the surrounding road, are acquired using sensors, and then integrated and analyzed via a network. A variety of systems are currently reaching the practical level such as: systems for automatically performing emergency notification in case of an accident, telematics insurance where premiums vary depending on driving record, and systems for tracking the position of a vehicle when it has been stolen.

[8] Abbreviation of "Mobility as a Service." A new concept of mobility using ICT, in which all mobility aside from one's own car is regarded as a single service to achieve seamless connection using cloud-based means of transportation.

[9] A thought experiment asking whether it is permissible to sacrifice B to save A. These sorts of ethical questions are attracting attention again due to the advent of self-driving vehicles.

Information previously lost with digital

The world is brimming with various forms of information that cannot be grasped with human senses. Some of this information is already being captured with a wide range of sensors and put to use. However, a very interesting world is likely to emerge if IOWN can incorporate information which previously could not be captured as data.

Consider film cameras as an example. People are taking another look at the value of film cameras, like the Cheki instant camera. Whereas a digital camera converts light entering through the lens into electrical signals, a film camera captures light directly, and the image is revealed through a chemical reaction.

When light is converted to an electrical signal in a digital camera, information not detectable by the human eye is stripped off as unneeded. But in a film camera, various types of information are imprinted, without being selected for the human eye. There is significance in focusing on non-visual light and signals if the images captured by a film camera have a character or depth unique to a film photography, or pick up some other distinctive element. This too will likely connect with natural technology.

In recent years, a similar approach has been apparent in the digital cameras installed in smartphones. For example, three cameras with different focal lengths are sometimes installed, so the user can choose instantly whether to take a telephoto, wide-angle, or superwide-angle photo, or take clear photos in the darkness. Approaches like this allow users to enjoy photos that were previously difficult to shoot. There is also technology in which five camera sensors capture the subject at different exposures, and the most beautiful possible photo is composed through computer processing. These are further examples where information not captured by the human eye is sensed and used.

If we naturally capture, and carefully pick up on, things like the vision of bees, the olfactory abilities of gorillas or dogs, or the hearing of bats, we will be able to greatly extend our five senses of human beings. Of course, incorporating every kind of information, including the natural world, or even attempting to do so, has previously been impossible due to limitations of the network environment. However, IOWN transmission technology will make this possible. Technology previously thought to be impossible will undoubtedly become real.

Eliminating the need for device operation

Finally, as an example of natural technology, let's consider future trends relating to smartphones. Smartphone applications allow us to do things that were impossible with previous telephones. On the other hand, there are more than a few people who feel stress in managing applications. To skillfully use devices at a high level, and with greater convenience, one must have the corresponding knowledge, techniques, and literacy. The digital divide—conspicuous especially among the elderly who are not used to digital equipment—is a serious issue.

One future form of user interfaces proposed by NTT, *Point of Atmosphere(PoA)*, aims to realize a natural world, of harmony between people and the environment, that does not require awareness of each device in one's daily life. Here, there are no longer any terminals. Systems understand, through environmental cues, the behavior, intentions, and feelings of people in each situation, and support thinking and decision-making through active encouragement.

If you get up to leave a room, the system will understand your destination and tell you which train to ride. If you go on a trip, it will tell you what clothes to take for the climate of your destination. When you enter a restaurant, it will tell you what dishes and wines suit your taste. These are examples of situations in daily life, but, we envision a system where, for example, if you familiarize yourself with the "points" that serve as contact points with the network, like a bracelet, you can obtain needed information without operating individual devices.

2. From electronics to photonics

For IOWN, we are overcoming the limitations of the conventional Internet, and using new transmission technology to build an unprecedented networked world. Part of our aim is to provide clues to help solve the problem of energy consumption, which has become an issue for all of humanity in recent times. We believe that the main keys for realizing this lie in converting from electronics to photonics, and integrating electronics and photonics.

At present, NTT is engaged in research to incorporate optical technology into integrated circuits, and if this is successful, it will be possible to incorporate optics into all types of transmission, short- to long-distance, as well as into computational processing.

Research results serving as a touchstone for this were published in the April 15, 2019 issue of the British science journal Nature Photonics. We believe this paper, describing an optical transistor that operates with the world's lowest energy consumption, has great value. This type of technology—partially integrating optical technology into electronic circuits—has been studied for more than 20 years at the NTT Basic Research Laboratories, but it has not been established as practical technology due to issues of large size and high energy consumption. In our recent research results, we were able to reduce power consumption by 94% compared to previous results, and thus the potential for practical application has improved.

This technology will support Digital Twin Computing, in which analysis and feedback processing are performed in real-time. If practical implementation is achieved, it will support services requiring mission-critical large-scale calculation where delay times cannot be tolerated, e.g., services such as MaaS and connected cars.

If this sort of photonics technology becomes practical, new principles will be established to solve problems of information processing capacity and power consumption, and this may be a breakthrough capable of easing concerns about the breakdown of Moore's law.

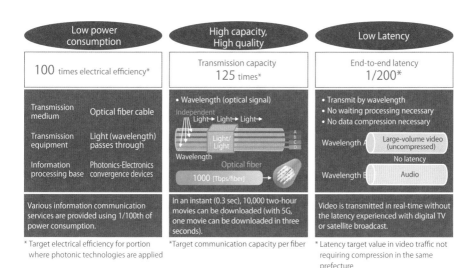

Low power consumption		High capacity, High quality	Low Latency
100 times electrical efficiency*		Transmission capacity 125 times*	End-to-end latency $1/200$*
Transmission medium	Optical fiber cable	• Wavelength (optical signal) Independent Light→ Light→ Light→ Light/ Light Wavelength Optical fiber 1000 [Tbps/fiber]	• Transmit by wavelength • No waiting processing necessary • No data compression necessary
Transmission equipment	Light (wavelength) passes through		Wavelength A Large-volume video (uncompressed) No latency
Information processing base	Photonics-Electronics convergence devices		Wavelength B Audio
Various information communication services are provided using 1/100th of power consumption.		In an instant (0.3 sec), 10,000 two-hour movies can be downloaded (with 5G, one movie can be downloaded in three seconds).	Video is transmitted in real-time without the latency experienced with digital TV or satellite broadcast.
* Target electrical efficiency for portion where photonic technologies are applied		*Target communication capacity per fiber	* Latency target value in video traffic not requiring compression in the same prefecture

Fig.4: ALL-Photonics for low power consumption, high quality, and low latency

5.　3 elements of IOWN

At present, IOWN is composed of the following three technology elements **(Fig. 5)**:

❶ All-Photonics Network (APN)
❷ Digital Twin Computing (DTC)
❸ Cognitive Foundation (CF)

❶ All-Photonics Network (APN) applies to all devices connected to the network. The idea is to realize end-to-end optical transmission by switching over from conventional electronics to photonics, i.e., shifting to optical technology in all information transmission (short to long range) and relay processing. In this way, we will break through the limitations of current Internet technology, and achieve high-quality, high-capacity, low-latency transmission with incredibly low power

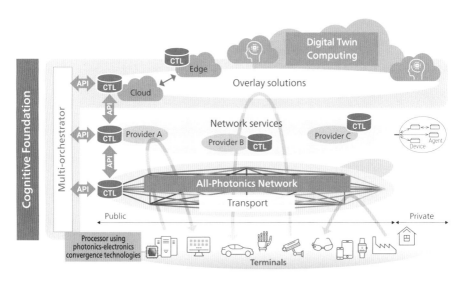

Fig.5: Composition of IOWN

consumption. We will also examine optimal connection of the APN with 5G (a 5th generation mobile communication system that started pre-service in 2019), and future wireless communication systems such as Beyond 5G.

❷ "Digital Twin Computing" (DTC) refers to technology for realistically recreating, in cyberspace, the objects, people, and other entities that make up the real world—thereby enabling sophisticated simulations that recombine those elements. The term "digital twin" was coined by the American Defense Advanced Research Projects Agency (DARPA). The idea involves recreating the phenomena of the real world on computers, i.e., creating "digital twins." DTC extends the available combinations and applicable scope of the basic digital twin concept. The aim is not to turn real space into data and copy it into cyberspace; rather it is to offer innovative services by modeling the real world at a high level, and achieving cross-over through interactive computational processing between diverse models. This corresponds to the "brain" of natural technology.

Here, the items needed for digital recreation of an object are selected from various sensing devices, and transmission within the network must be done at the necessary quality (e.g., signal quality and transmission delay). Also, today's people must operate their device to properly select 4G or WiFi for wireless access, and there is a need for the optimal wireless system to be assigned naturally without the user being aware of it.

❸ "Cognitive Foundation" is a system for allocating finite ICT resources while harmonizing to achieve an overall optimum, and distributing the necessary information within the network.

6. History of NTT's ICT concepts — From multimedia to multivalue

Taking ISDN—realized 10 years after the INS initiative in 1979—as a departure point, NTT has gradually progressed in its efforts to achieve multimedia communication linking information and networks (Fig. 6). Today, these efforts have borne fruit as more secure and convenient networks based on conversion

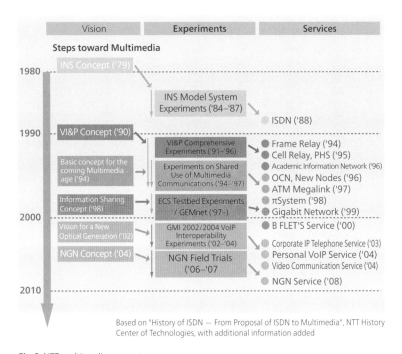

Vision	Experiments	Services

Steps toward Multimedia

1980 — INS Concept ('79)

INS Model System Experiments ('84–'87)
● ISDN ('88)

1990 — VI&P Concept ('90)

VI&P Comprehensive Experiments ('91–'96) → ● Frame Relay ('94)
● Cell Relay, PHS ('95)

Basic concept for the coming Multimedia age ('94) → Experiments on Shared Use of Multimedia Communications ('94–'97)
● Academic Information Network ('96)
● OCN, New Nodes ('96)
● ATM Megalink ('97)

Information Sharing Concept ('98) → ECS Testbed Experiments / GEMnet ('97–)
● πSystem ('98)
→ ● Gigabit Network ('99)

2000

Vision for a New Optical Generation ('02) → GMI 2002/2004 VoIP Interoperability Experiments ('02–'04)
● B FLET'S Service ('00)
● Corporate IP Telephone Service ('03)

NGN Concept ('04) → NGN Field Trials ('06–'07)
● Personal VoIP Service ('04)
● Video Communication Service ('04)
● NGN Service ('08)

2010

Based on "History of ISDN — From Proposal of ISDN to Multimedia", NTT History Center of Technologies, with additional information added

Fig.6: NTT multimedia concepts

of communications to digital and optical technology.

IOWN aims to evolve toward a "multivalue" stance—accepting and supporting the diverse values of society and the next era of networks. In the run up to 2030, ten years from now, we plan to exploit the technology we have accumulated thus far, and advance R&D across a broad domain, ranging from foundations to applications. At the same time, we aim to realize the IOWN initiative while collaborating with global partners and experts in a wide range of research and technology fields.

Chapter 2

11 Technologies and 3 Elements that Drive IOWN

Compiled by the NTT Research and Development Planning Department

1.   11 technologies that will advance together with IOWN

The world is currently undergoing radical changes. This is mainly due to the progress and dissemination of ICT in recent years, and innovation driven by various state-of-the-art technologies. The Internet in particular has dramatically lowered the transaction costs of information and goods, removed physical constraints, made the world smaller in one stroke, and caused major changes in the state of society.

On the other hand, the limits of this process have come into view. Due to the breakdown of Moore's law, the outbreak of global social problems (e.g., those addressed by SDGs), the strengths and weaknesses of digital society, and global fragmentation, we need to build a Smart World to succeed today's society, and a new future with emphasis on the sustainable development of humankind. That is the role of IOWN.

As technologies to help build this Smart World, we have identified the 11 technologies indicated below [1]:

01) Artificial Intelligence (AI)
02) Virtual Reality / Augmented Reality (VR/AR)
03) Human-Machine Interface (HMI)
04) Cybersecurity
05) Information Processing Infrastructure
06) Networks
07) Energy
08) Quantum Computing
09) Biotechnology / Medical Care
10) Advanced Materials
11) Additive Manufacturing

These key technologies are essential for solving the social issues the world faces, and will contribute to the sustainable development of humanity. At the same time, they are technologies that will both advance IOWN, and evolve due to IOWN. That is, these 11 technologies and IOWN will accelerate each other's development in a complementary fashion, and solve the issues confronting humanity. The world of IOWN will be supported by advanced technology in diverse domains, and IOWN will shine even brighter by supporting advanced technologies.

Now, let's look at profiles of these 11 technologies for supporting a Smart World.

[1] We conducted a survey on social trends in 8 globally important industries (real estate and construction; retail; manufacturing; healthcare; automobiles and mobility; finance; energy; and agriculture, forestry and fisheries), identified technologies supporting social trends, and pinpointed 11 technologies by classifying technologies while taking a broad overview of these 8 industries.

01   Artificial Intelligence (AI)

Artificial intelligence (AI) research is currently accelerating all over the world, amidst progress in big data analysis, machine learning, and deep learning, and headway is already being made toward practical application in many domains. As is well known, results have been achieved that surpass human abilities in specific areas such as media processing. However, today's AI has still not reached the desired level in terms of approaching versatile, general human intelligence.

It's difficult to rival human beings in a single leap, and the next likely step is human-centered AI that accepts the diverse values of individuals. AI needs to do more than just deduce the answers to problems it is given. We need AI grounded in diverse human values—AI that understands things like the background of ideas and situations, and works closely with people while always taking account of individual values.

02   Virtual Reality / Augmented Reality (VR/AR)

Various applications of virtual reality / augmented reality have already been announced, and this technology is expected to grow rapidly in the future. Practical systems have been around for a long time, in areas like driving simulators, and in recent years the technology has been enthusiastically adopted in sectors like the entertain-ment industry. Both B2B and B2C businesses are examining practical use in various areas, and moves toward social implementation have begun.

To enable use in more diverse situations, it will be crucial to "intensify the sense of presence" and more faithfully replicate the real presence of objects. Higher pro-cessing speed and prevention of simulation sickness will also need to be sought. By overcoming these problems, and refining technology to integrate elements such as vision, audition, and tactile senses in a more sophisticated way, the sense of presence will reach a new level even closer to reality.

03 Human-Machine Interface (HMI)

Human-machine interfaces (HMI) are a technology/system for mutual exchange of information between human beings and machines (artifacts). The concept of HMI has broadened and deepened due to recent progress in related technologies such as VR/AR, cognitive science, neuroscience, and robotics. If VR/AR is regarded as an extension of human senses, HMI can be regarded as an extension of human "sensation and movement" through bidirectional interaction between humans and machines.

Continued research will be needed to further deepen the HMI concept—naturally extending the bodies of human beings, and understanding them at a deeper level, so humans can use devices and robotics without being explicitly aware of it.

04 Cybersecurity

With the progress of ICT, cyberattacks have become increasingly sophisticated, and there is a risk of attacks becoming larger in scale. There has been an explosive increase in devices connected to the Internet due to the dissemination of IoT, and this creates a broader range of objects to attack. Attacks on highly developed networks and IoT devices have the potential of causing major incidents in the real world.

In order to minimize damage, it is crucial to not only withstand attack, but also to deploy preventive security measures. Preventive defense involves, for example, early understanding and prediction of attack techniques and the damage situation through approaches like machine learning. This makes it possible to deal more nimbly with attacks, e.g., mitigating damage and preparing for attacks before they occur. The threat of cyberattacks is evolving every day. Dealing with this threat will require research as well as improved preventive measures. This research will need to broaden the scope of defense to elements of the real world like IoT and mobility, and examine encryption technology for continuing to protect information in any future computing environment.

05 Information Processing Infrastructure

Running sophisticated AI and implementing IoT will require an information processing infrastructure for processing large amounts of information in real time. Research and development is accelerating in this field, and the AI chip market in particular is expected to see rapid growth.

When various types of equipment incorporate AI devices—i.e., devices which deal with the real world, recognize situations, and respond autonomously—a need will arise to convert all sorts of real world events into data in real time, distribute that data, and respond in a timely fashion throughout the society as a whole, without the data being confined within individual devices. For example, if there is an obstruction on the road, and a self-driving vehicle avoids it, the vehicle should simultaneously notify its surroundings of the presence of the obstruction, and other vehicles that have received that notification should change their path.

Realizing a world like this will require infrastructure to convert events in the real world (states of things, occurrence of events, etc.) to information, and distribute that information. For this reason, we will need a high-speed, ultra-low-latency network, and an information processing infrastructure with dramatically improved calculation capability and efficiency.

06 Networks

As people in countries throughout the world are increasingly connected by networks, and connected devices—such as IoT equipment and connected cars—proliferate, the requirements on networks will become increasingly stringent. In particular, there will be a need for increased network speed and greater sophistication of network control, and R&D in these areas is moving forward all over the world.

In terms of network speed, development is focusing on sophisticated optical transmission technology and wireless transmission technology, including spatial multiplexing technology. In the area of network control, various approaches are being tried, including edge computing, network virtualization, network optimization, and automatic operation using AI.

Innovative networks created in this way will connect people and devices everywhere, and serve as essential infrastructure for the smart, affluent society of the future.

07 Energy

As the world population increases and ICT continues to grow, energy demand will increase. It will be essential to optimize power usage in society as a whole, by, for example, using renewable power to protect the global environment, and smart grids to coordinate electric power supply and demand. In Japan, efforts are underway to achieve local production/consumption of electric power, and electricity self-sufficiency. These efforts include interconnecting distributed power supplies like solar panels and storage batteries (including electric vehicles), and trading power between individuals and businesses.

Going forward, technology will be needed to optimize supply and demand within the energy distribution infrastructure, using real-time energy supply and demand matching for "smart" distribution of energy. Technology for efficiently converting energy to electricity, and research on basic technologies for green energy, like artificial photosynthesis, will play an important supporting role. Through these approaches, we can likely remake energy as social infrastructure.

08 Quantum Computing

There are high expectations on quantum computers, as a means of solving previously insoluble problems, because they allow information processing using quantum-mechanical properties.

Efforts aiming to achieve practical use are accelerating in various industries, targeting applications related to optimization problems in particular. Quantum computers are one solution for problems which take an inordinate amount of time to produce an answer, due to the tremendous amount of calculation required when using a conventional computer.

A quantum computer can immediately determine optimal solutions from a huge number of alternatives. This is useful, for example, in optimization of logistics, inventory, or shop locations; optimization of mobility or human traffic flow routes; and optimization of the arrangement of functions within a city.

09 Biotechnology / Medical Care

Deeper understanding of bio-information has been achieved due to the development of biology, chemistry, and medicine, combined with the evolution of technology. This has resulted in rapid progress in the domains of biotechnology and healthcare. Areas such as molecular-scale manipulation and multidimensionality of molecule design have been a recent focus of attention in biotechnology research. In the area of high-precision, multi-dimensional design at the cellular level, one important topic is regenerative medicine using iPS cells.

On the other hand, research results from biotechnology have come into use not only in medicine and healthcare—and the agriculture, forestry, and fisheries industries which supply the food essential for maintaining human health—but also in the domain of information and communications. For example, biometric authentication technology is useful in verifying identity for financial transactions, and is becoming the mainstream approach for smartphone authentication. Going forward, use of this technology across different fields will likely progress, including use in information and communications.

10 Advanced Materials

Research on new advanced materials, which surpass the concepts of conventional materials, has made progress in recent years, with nanomaterials and biomaterials at the top of the list. Development of these advanced materials will contribute to the improvement of QOL (Quality Of Life) in medicine and other fields, and will also serve as a foundation for innovation in manufacturing and other industries.

What is needed at present is accelerated material development—narrowing down combinations of materials and molecules, through simulation and machine learning, to hold down costs in terms of both money time. Another important topic in the medical field is personalization of function for realizing treatments and drug design suited to individuals. The aim here is form and function with outstanding biocompatibility in areas such as molecular targeted therapy and cell therapy.

11 Additive Manufacturing

Additive manufacturing, exemplified by 3D printing, is a technology for creating a desired finished component by layering or adding materials, as the name suggests. Because it can loosen design constraints and incorporate diverse functions at low cost, this technology is indispensable for creating a Smart World.

The technology is still in its embryonic phase, with many issues still to be solved such as diversification of materials, increased layering speed, and improved precision. Further innovation will lift expectations in areas such as practical 4D printing, where shapes can vary with the passage of time, and applications to the bioprinting field.

2. 3 elements of IOWN

In the section above, we provided a rough overview of the 11 technologies that will help build a Smart World. Next, we will explain in detail the 3 elements comprising IOWN that were mentioned in Chapter 1:

❶ All-Photonics Network (APN)
❷ Digital Twin Computing (DTC)
❸ Cognitive Foundation (CF)

These 3 elements correspond, respectively, to the key functions of transmission, computation, and allocation of ICT resources for connecting transmission/computation. They also correspond to the main functions constituting the network and information processing infrastructure. As in the case of cyberphysical systems realized using DTC, development of a computing environment re-

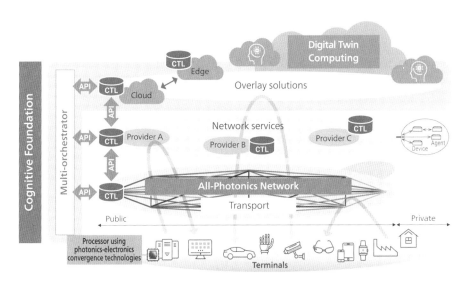

Fig.1: Composition of IOWN

quiring large-scale computing and high-capacity communications is difficult to achieve without low-power-consumption, high-capacity, and low-latency transmission provided APN, and optimal allocation of all IT resources provided by CF. That is, in the IOWN initiative, technology development for APN, DTC, and CF must not be advanced toward individual goals. R&D must aim to realize an innovative vision in which the three elements are unified.

Therefore, at NTT, we will promote wide-ranging technology development across multiple fields, based on previous R&D results.

The respective roles of the 3 elements of IOWN are as follows.

1 All-Photonics Network (APN)

An All-Photonics Network is a network in which all communication is carried by light from the sender to the receiver. In today's Internet circuits using fiber optic transmission, conversion between optical and electrical signals must be done several times via routers. The aim with APN, on the other hand, is to achieve communication with optical signals alone, without any mediation by electrical signals.

Realizing APN will require design of function-specific topologies tailored to the various communication requirements of different services, technology to optimally assign finite wavelengths, and dynamic wavelength path connection technology to accommodate all-photonic communication in finite optical fiber. It will also be important to increase capacity per fiber, and conduct R&D on new optical devices such as optical multiplexing, switching, and wavelength conversion elements. Ultra-high-speed wireless technology for achieving higher speed, including wireless access, will also be indispensable.

■ Benefits of an All-Photonics Network

By connecting all communications devices via light instead of electrical signals, and using more wavelengths, the network will achieve high-capacity, ultra-low-latency transmission far superior to current networks. Innovations to increase capacity will enable sending/receiving of information previously lost at the stage of digital conversion, in a high-definition condition closer to the original signal. We can also imagine use of information imperceptible to human beings, e.g., non-visible light in the case of video communication. The ability to widely transmit high-capacity information will provide benefits for all sensor networks.

If a different wavelength is assigned to each type of information in fiber optic transmission, it will be possible to send multiple types of information simultaneously with ultra-low-latency. One example might be ultra-low-delay interactive exchanges, while sending high-definition images on multiple channels. Practical use in settings where communications quality is a critical requirement, such as remote surgery, will also come into view.

2 Digital Twin Computing (DTC)

Digital Twin Computing is a development of the previous digital twin concept—a new computing paradigm enabling free interaction in cyberspace, beyond mere recreation of the real world. This will involve combining diverse types of digital twins and employing various types of calculation.

A digital twin is an accurate representation—in cyberspace—of the form, state, function, and other properties of an object in the real world such as a production machine in a factory, an aircraft engine, or an automobile. Accurate information representing people, such as MRI data or CT images in the medical field, are also an existing form of digital twins. At present, however, digital twins are specialized for recreating objects in the real world (e.g., things and people) for specific purposes.

Our aim is to foster free interaction between digital twins in cyberspace, and go beyond mere recreation of the real world. To achieve this, we will further develop previous digital twin concepts, combine diverse types of digital twins, and perform various types of calculation. For example, by mutually exchanging, integrating, and copying data and models constituting digital twins—of both people and things—it should be possible to build and simulate intricate, complex virtual societies in cyberspace, and provide feedback on the results to the real world.

Realizing DTC will require digitalizing the inner aspects of human beings, such as feelings, thoughts, and values, and developing technologies to carry out simulation by dynamically combining people (with such inner aspects) and things in the same space.

■ Benefits of Digital Twin Computing

Sophisticated interaction of physical space and cyberspace based on DTC is likely to result in solution of problems and provision of services, as indicated below.

① Design of future cities

DTC will enable urban design of a sort that does not exist today. For example, when creating new cities, it will be possible to simulate beforehand the optimal city size and structure. This will be done by reconstructing, in a virtual world, the geographical conditions and climate of the prospective location, and arranging objects—such as houses, buildings, and electric power—synthesized by combining digital twins of infrastructure in existing cities. By deploying people there as digital twins, it will be possible to recreate human-city interactions involving lifestyles and culture, and simulate even the social activities that will animate the city.

② Predicting the future of real-world society including people

The digital twins of people in DTC are modeled with individuality in terms of thinking, decision-making, and behavior processes. Therefore, if a virtual society is constructed by bringing together multiple digital twins, it will be possible to simulate the daily lives of people in the virtual society and predict the future. Unlike simulations using uniform, statistically-modeled people, this approach enables more precise simulations reflecting the thoughts and behaviors of diverse individuals. The results of simulating an entire virtual society can be used for more than just organizational and social decision-making. For example, simulation results from the micro-perspective of individuals can be used to determine one's own conduct.

3 Cognitive Foundation (CF)

Cognitive Foundation is a framework enabling unified construction, setting, management, and operation of the cloud, networks, and user ICT resources. Previously, these ICT resources have been siloed, and managed/operated individually, and this has been a major obstacle to realizing high-level decentralized cooperation in edge computing or the hybrid cloud.

Realizing CF will require user-friendly operation technology enabling complete automation and autonomy—so the user never needs to be aware of the network—as well as technology for achieving autonomy and complete automation of operation to deal with cyberattacks.

It will be essential to improve the level of the system as a whole, both wired and wireless. Due to the use of proactive area realization and multiple wireless cooperation technologies to suit the user situation, it will be essential to realize a communications environment with no user awareness of the wireless network. At NTT, we use "Cradio" as a generic term for wireless control technology in CF, and we will accelerate our R&D in this area.

■ Benefits of Cognitive Foundation

With CF, rapid deployment/optimization of ICT resources and optimization of application configuration (automatic design, autonomous operation) are achieved through cooperation between ICT resources and the Multi-orchestrator. Furthermore, more sophisticated management and operation are achieved by providing intelligent functions. As a result, CF contributes to digital transformation of one's own work, and improved operational efficiency of businesses that are the middle B in the B2B2X model by using the efficient smart operation without necessity of the manual operation. It also achieves improved convenience in areas such as providing speedy services to end users.

These technologies will not be something which users who routinely come into contact with IOWN are aware of. However, they will be extremely important for providing holistically optimal services, and building the IOWN world, in which an extremely vast amount of information is gathered, transmitted, processed, and provided as feedback.

3.　11 technologies and 3 elements of IOWN

We believe the eleven technologies and three elements of IOWN, outlined thus far, will constitute the world of IOWN not through hierarchical relationships, but through complex entanglement as seen in the phenomenon of nesting.

Now, let's consider how the 11 technologies and 3 elements of IOWN—areas where R&D is expected to heat up worldwide going forward—will connect, and how to accelerate both the technologies and elements while sparking interaction between the two. Among the 11 technologies, there are some which are closely connected to APN, DTC, or CF, respectively, and some which are connected to IOWN as a whole. Let's look at this in more detail.

■ Technologies for advancing All-Photonic Network

First, let's take a broad overview of technologies for configuring and accelerating APN. To realize APN, an important key phrase is "photonics-electronics convergence," i.e., employing light for information processing and all data transmission, from short to long distance. The basic technologies needed for this are steadily coming to fruition (Part 2, CS9: AI Optical interconnect technology, CS11: Integrated optical front-end device technology, CS15: Nanophotonic device technology, etc.). The keys in terms of technology will naturally be "06) Networks" technology, as well as "10) Advanced Materials" and "11) Additive Manufacturing" for developing innovative devices and elements for an all-photonic system.

■ Technologies for advancing Digital Twin Computing

Next, let's consider the technologies that will constitute DTC and the applications that will use it. In DTC, things and people in the real world are modeled at a high level. This will provide the ability to perform ultra-large-scale, wide-ranging, high-speed, high-precision prediction based on the real world. To achieve this, extensive use of "01) AI" technology will naturally be indispensable. Also, "02) VR/AR" and "03) HMI" technology will be essential as interface technology for presenting information to people in a natural way. There may also be applications of technology related to "08) Quantum Computing" in providing innovative processing capabilities.

■ Technologies for advancing Cognitive Foundation

In CF, the object of management is not just resources of the transmission network; it is all resources for achieving any intended processing, including sensors and IoT devices that gather information, and computing resources for analysis and processing of the gathered information. Integrated management is carried out while striving for overall optimization. To achieve overall optimization of extremely large-scale resources, the key will be realizing complete automation and autonomous operation without any human intervention. Integration of "01) AI" and "06) Networks" will likely be a major theme.

■ Technologies for advancing IOWN as a whole

Finally, let's consider technologies for advancing IOWN as a whole, across all 3 elements.

Secure, high-speed, large-scale, high-precision information processing, based on the two technologies of "04) Cybersecurity" and "05) Information Processing Infrastructure", will be the essential core for creating IOWN as "information environment = place", and this will evolve in close relation with all three elements of IOWN.

■ Technologies advanced by IOWN

Next, let's consider technologies that will be accelerated by the realization of IOWN.

In the field of "07) Energy" more sophisticated simulations are possible with extensions of current technology, but in the world realized by IOWN, far more extensive, high-precision simulations will likely be possible. Also, in a world where sophisticated prediction technologies have been established based on DTC, we are likely to see new approaches in the "09) Biotechnology / Medical Care" field, including precision and personalized medicine employing AI analysis of personal medical data, as well as smart healthcare for ascertaining, in real time, the physical and mental condition of people in daily life. In these fields, IOWN has the potential of making major contributions to the dramatic improvement of prediction precision.

The 11 technologies attracting attention as technical trends will evolve further in the future, supporting and becoming more pervasive in the daily lives of people. Through IOWN, we hope to link these technologies with our new "information environment = place" approach, and develop them further. We are convinced this will help solve the various issues humanity faces in modern society.

Thus far, we have outlined the 11 technology trends and three elements which will shape the world of IOWN, and their relationships. The eleven technologies are enhancing our lives as they are refined at a dizzying pace by companies and research centers inside and outside Japan. As benefits of their evolution, they will help create the world of IOWN, and IOWN will further accelerate their development.

Chapter 3

The Future Society Created by the IOWN Initiative

Motoyuki Ii
Senior Executive Vice President and Member of the Board, NTT Corporation

1.　Aims of the IOWN initiative

From communication to prediction of the future

The IOWN Initiative was launched in 2019 with an eye toward 2030, ten years later. In this section, I'd like to consider what sort of society we will see in the near future.

Looking back on previous progress in information and communications technology, the aim in every case—be it the telephone, ISDN, or fiber optic communication—has been improvement in communication. First it was voice; then it became possible to send and receive text, images and videos. The value of communication has grown as definition and capacity have increased. Before long, an IoT framework came into use in industry and daily life—a framework going beyond person-to-person communication, and using sensor-gathered data for person-to-thing and thing-to-thing communication. Nevertheless, the essence of the technology thus far has been communication.

The aim of IOWN is a new stage, unlike anything before. The key will be predicting the future. Taking the existing communication-centered as a foundation, we will create new value in predicting the future. If accurate prediction can be achieved, we can respond accordingly. This will be tantamount to "changing the future."

Changing the future — How will people's values and sense of well-being change when that becomes possible?

Extremely quick, accurate predictions of the future will change the real world

Comparatively simple forms of predicting the future are possible even with today's systems. The distinguishing feature of predictions by IOWN, however, will be accuracy and speed an order of magnitude greater.

For example, human traffic flow prediction systems predict how many people will move and in what way in public places. These systems are used at shopping centers and other locations to control air conditioning and detect danger. They gather data—sent in from sensors installed at various points—on number of people, room temperature, energy consumption, and other parameters. Then they perform calculations, and output prediction values. IOWN, however, will enable more precise air-conditioning control based on faster, more accurate predictions. Energy-saving effects will also be enhanced. It will be possible to quickly predict things like detection of suspicious objects or dangerous locations, and immediately detect people who are acting suspiciously. Dealing with situations proactively, based on predictions, will enhance effectiveness in preventing accidents and incidents.

Sophisticated prediction of the future using biodata is a possibility in the field of medicine and healthcare. By combining daily biodata such as body temperature, blood pressure, and pulse rate with a person's previous history of illness, and then performing computation, it will be possible to accurately predict what sort of diseases the person will experience and when. This will enable prevention of illness before it occurs, and quick response when it does. If trends in illnesses and infectious diseases can be ascertained for society as whole, it will be possible to develop drugs and treatments beforehand. However, adequate measures must be taken to protect personal information when employing this sort of prediction.

IOWN can also be used for self-driving vehicles which require accurate, split-second judgment. A vast amount of information is produced by cameras installed in vehicles and on roadsides, and by quickly transmitting this information and processing it at high speed, it will be possible to quickly produce accurate traffic jam predictions and deduce alternate route information. Furthermore, intelligent control of traffic throughout an entire region will help prevent traffic jams and accidents from occurring. Accurate prediction will allow us to change the future.

Better options through prediction of the future in cyberspace

Digital Twin Computing (DTC) will enable prediction of the future in cyberspace before things happen in reality. This is difficult within the framework of today's computational software, but the new DTC approach will enable integration of the world of things and the world of people at the virtual, cyber level. We will be able to obtain solutions for a future point in time by combining various elements in cyberspace, and then performing computational processing at high speed.

For example, if you want to discuss some issue, you might input the discussion participants and theme, and the system may be able to immediately determine the conclusion. If it would take one hour to reach the same conclusion, were the discussion to be actually held, then producing the conclusion immediately at the cyberlevel means it is possible to predict the future one hour later.

Naturally, there may be more than one solution. If the conditions are changed, or the ideas of the people in cyberspace are adjusted, then one can obtain multiple solutions for each set of conditions. Humans can then choose the best solution from those alternatives. Until there is greater progress in areas like brain science, it will be difficult to recreate full-fledged human thought in cyberspace. However, if we can recreate and use thoughts, feelings, and values, at least partially, this may be useful for decision-making in various settings.

2.   Open innovation for the world

Commencing efforts to realize our vision

Realizing our vision will require new thinking and technology not available with today's ICT technology. By using Cognitive Foundation (CF) to optimally control All-Photonics Technology (APN) for transmission, and Digital Twin Computing (DTC) for producing new services, NTT will realize infrastructure technology with unprecedented low power consumption, high capacity, low latency, and high-level information processing, and thereby drive the IOWN concept. Low power consumption is extremely important considering limited availability of energies on global scale.

NTT has 12 laboratories at work not only in the field of information and communications technology, but also in a wide range of other fields including brain science. In each field, we emphasize applied as well as basic research, and notable results have been achieved in areas like optical technology and optical Ising machines. The IOWN initiative will require further advances in this sort of research, and naturally it will be impossible for NTT to achieve this alone. We will seek out a broad range of partners, world-wide, and move forward through open innovation.

A number of new projects are expected to begin in the near future, but here I would like to present some services we are already working on in collaboration with our partners.

Smart Cities for proactively preventing crime and accidents

Since the fall of 2018, NTT has been engaged in a Smart City initiative together with the city of Las Vegas (Nevada, U.S.). A system called "Public Safety" for ensuring municipal safety gathers data from crime-prevention cameras citywide, and uses AI to determine the degree of danger from the obtained images, such as whether people are carrying dangerous items, or there are places where crowds are gathering. Crowd congestion is also included in the check items because it is empirically true that the probability of an incident increases with the number of people.

At present, this system is still under development, and the police deal with situations that arise using human man-power. However, if this system actually comes into service, it will be possible to provide the needed amount of security staffing exactly when it is needed. This will help proactively prevent incidents before they happen, and make cities safer. There are no existing examples of security systems that prevent crime over the extensive area of an entire city, and if such a system is realized, it will definitely generate a positive response.

To prevent crime, NTT offers an AI service for shoplifting prevention called *AI Guardman*® [1], and it has already achieved results at locations like the cosmetics departments of volume retailers. Cosmetics are small and expensive, so they tend to be involved in a large number of shoplifting cases and incur high costs. With this service, an AI camera is placed on the scene, and it judges danger based on the behavior of a person who has picked up a cosmetics product.

In this system, a camera is equipped with an AI chip for judgment. This judgement is carried out by regularly receiving and updating the latest suspicious behavior pattern file from the AI cloud, and performing video analysis within the camera itself. Behaviors characteristic of suspicious people are detected and incorporated into the judgment process, and if the person under observation is determined to be suspicious, a nearby salesperson is notified and staff are deployed. Staff can then take steps like speaking to the suspicious person to proactively prevent crime. Previously, judgment was impossible until the person had actually taken a product out of the store without paying at the register, and that put a heavy burden on the store. Stores which have adopted this system have reduced losses compared to the previous system relying on human man-power, and they say there are major advantages in terms of both costs and alleviating burdens on people. However, the behavioral characteristics peculiar to suspicious people differ depending on country and culture. This solution, which proactively prevents crimes and incidents, is one example of imagining a world that makes active use of IOWN.

[1] Registered trademark of Earth Eyes Co., Ltd.

The future of primary industry, made possible by semi-automation with IT

In recent years, there has been a remarkable shift toward the use of IT in agriculture. Farmers are already acquiring data on crop growth and insect damage from drone videos, and putting this to use in areas like predicting harvest times and planning pesticide application. Here too, repeated accumulation of data using IOWN will reveal characteristics such as the growth of each variety or each subdivision of land, and this knowledge can be used to predict the future of farm produce. This approach can also be used in areas like rice transplanting and reaping, and when smart self-driving cultivators and harvesters arrive, this will free up agricultural workers from heavy agricultural labor.

At present, there are worries about a lack of successors in agriculture, and thus we believe agricultural IT will become increasingly important as a method for increasing yield. This method will not rely on the experience and intuition of proficient farmers, as in the conventional approach. Other primary industries such as stockbreeding, forestry, and fishing face similar problems. We are moving into an era where past empirical knowledge will no longer be reliable due to the rapid pace of climate change. If accurate predictions and decisions can be made using IOWN, this will be very beneficial for primary industries.

3. Realizing the society of the future

Weighing the risks of technology development

How will society change as IOWN-based prediction of the future moves toward realization? AI based on machine learning and deep learning has been in the limelight since 2012, and its dissemination has drawn attention to the problem of "the singularity" as a real possibility. In his book Homo Deus (Vintage, 2017), the Israeli historian Yuval Noah Harari claims that humans—equipped with science and technology—will introduce changes outside the framework of natural evolution. He warns that a stratified society beyond imagination may arise due to such changes.

New technology is always developed in pursuit of human happiness, but when it reaches the practical level, technology always carries a risk of abuse, and disadvantages to go with its advantages. In addition, the stronger the impact of a technology, the greater the damage it may inflict on society.

Gene editing and iPS stem cell technologies are being developed to contribute to medicine, fully cure diseases that previously could not be treated, and save the lives of people who cannot be saved with conventional medicine. On the other hand, the same technology enables applications that are not ethically permissible. No major problems have yet occurred, but there is no social consensus in place, and regulations have yet to be established. Under these conditions, it would not be surprising to see unforeseen mishaps in the future.

The same holds for IOWN. When accurate, fast prediction of the future becomes a reality, no one knows what will happen. If someone abuses future prediction to obtain self-interested benefits, it may be impossible to turn the clock back. To prevent such a situation from occurring, ethics will have to be extensively discussed in parallel with technology development. We will have to consider measures for dealing with such problems.

How will we live in the reality of cyberspace?

If another "I", corresponding to a real-world person, is created in cyberspace using DTC, it will be possible for cyberspace reality and real-world reality to intertwine with each other. Here too, we will need to imagine scenarios—what sort of problems may arise, under what circumstances, and what measures can be taken to address those problems.

Even today, there are problems like children addicted to computer games, and bullying on social media. People have avatars of themselves in a game, and as they cooperate and fight with the avatars of other players, the game becomes the "place where they live." For some people, this interferes with daily life. They don't go to school, don't eat with their family, and shut themselves away from society. Some young people commit suicide due to bullying on social media.

These issues go beyond the world of games and social media. When alternate realities are provided commercially as experience services in cyberspace, what sort of values will users of the services rely on as they live in two different realities, the real world and the cyber world? Considering both the positive and negative aspects will be essential.

Importance of discussion within a consortium

Like all previously developed technology, the world created by IOWN may have a negative as well as a positive side. In establishing a consortium, we want to create a forum for open discussion to not only promote technology development with our partners, but also deal with the socially complex, difficult problems that IOWN is likely to bring about.

Many of these problems will lie outside the domain of science and technology, and the discussion will need to involve experts in the humanities and social sciences, such as philosophers and sociologists. PR activities will also be essential for educating the broader society on the importance of these problems, and disseminating questions that people will be concerned about.

Juichi Yamagiwa, a world-class gorilla researcher and President of Kyoto University, has compared monkey society and gorilla society in his book Monkey-Like Trends in Human Society (in Japanese, published by Shueisha International, 2014). Whereas monkey societies always establish an alpha male, and emphasize vertical relationships, gorillas establish peaceful, egalitarian relationships without much of an order. In terms of evolutionary lineage, humans are closer to gorillas (anthropoid apes) than to monkeys. Therefore, it would be natural for humans to develop a society like gorillas. But what do we see in recent human societies? Disparities are growing, all over the world. President Yamagiwa is sounding the alarm bell—that humans may be shifting towards a monkey-like society.

When science and technology advance, human beings adopt the new innovations, and humans themselves are greatly changed as a result. Harari describes a future where humans change not only due to natural evolutionary processes, but also due to science and technology created by human beings. We are entering an era where we must seriously discuss technological and social problems—establishing measures to prevent abuse, and creating a society that is more "human" and less "monkey." The IOWN initiative calls on us to collaborate with a broad range of people in thinking about and creating such a future for humankind.

IOWN and 6 Industrial Use Cases

Compiled by the NTT Research and Development Planning Department

Use Case 1
MaaS + IOWN
Realizing the ultimate in smart mobility

For the future of mobility

Dissemination of self-driving vehicles—equipped with autonomous control systems connected to the network—is rapidly progressing in the world today. On the other hand, development of general support systems for transportation, using ICT as a foundation, are also moving forward on a large scale. The concept of MaaS (Mobility as a Service) is attracting particular attention.

MaaS is one target of mobility innovation for the future. MaaS is an ultra-smart transportation system using ICT in which all transportation services are seamlessly connected, and individual users can access optimal, safe mobility services suited to their moment-to-moment needs. In Japan, one high-priority measure in the "Future Investment Strategy 2018" is to create model cities and regions for new mobility services suited to user needs, and the Ministry of Land, Infrastructure, Transport and Tourism has commenced efforts in this area[1]. At present, efforts are being made across the boundaries of companies and industries, with the government and private sector working as a team, and creation of new systems for achieving MaaS is being examined at a rapid pace.

Let's imagine the future that will arise due to these efforts. Individuals will be able to engage in their regular patterns movement, like daily commuting and going to school, without paying attention to the specific mode of transport. Even in movement which arises sporadically in daily life, they will be able to select transportation modes most suitable at the moment, as they are dynamically suggested by the system. This may bring about a world where people can avoid traffic jams and packed trains, optimize energy consumption, and minimize the stress they face in moving around.

However, realizing this will not be easy. Transportation services will need to have not only the ability to immediately respond to people's diverse needs, as they change moment-to-moment, but also real-time sensing of the situation in the city and transportation network as a whole, integration of information, and optimization to stably maintain overall harmony. That is, it will be essential to have communications systems for gathering a vast amount of information at high speed, analyzing it in real-time, and supporting the stable operation of high-level cooperation/control systems. Extending current communications technologies will incur a major burden in many areas like data capacity, reliability, and energy consumption. Therefore this is a field requiring the realization of IOWN.

4 Use Cases

① The ultimate fail-safe

The term "fail-safe" refers to a technology or service that protects users from unforeseen circumstances that may arise in transportation, and changes a dangerous condition into a safe condition. Various approaches have already been tried, such as improving the safety equipment of vehicles themselves, but we envision fail-safe next-generation services making maximal use of a communications network based on IOWN.

One example is the realization of "public cooperative driving." This service strives to maximize safety of both individual drivers and the overall system. Through a high-speed optical network enabling all-photonics operation, integrated ICT resource allocation, and high-speed low-latency information processing, the system ascertains the relationships between vehicles and the overall transportation situation in an area, and makes decisions to promote public welfare and safety.

This service will also be effective in major disasters. For example, if a disaster like an earthquake occurs, there will be a severe traffic jam if everyone independently drives their own car, and the entire transportation network will be paralyzed. In MaaS realized through IOWN, the current state of the entire region is ascertained via sensing, simulation is carried out to avoid traffic jams, and navigation advice is provided to individual vehicles based on more accurate judgment. The guiding concept is to provide overall optimization while being tolerant of user diversity, and in line with that, the system has a bird's-eye view of the whole and the ability to make value judgments, and it encourages optimal behavior by each person.

② Proposing optimal transportation routes

At present (2019) companies both inside and outside Japan are working to develop flying cars. The "Roadmap toward Air Mobility Revolution[2]" of the Ministry of Economy, Trade and Industry advocates "starting the project in the mid 2020s, and shifting to practical application in the 2030s." In the near future, the routine transport modes of people will expand from two-dimensional routes on the ground to the three-dimensional space of the sky. When that happens, IOWN will provide high-reliability support for navigating in three-dimensional space—thanks to its high speed and low latency—and contribute to experiments on the service infrastructure for ensuring safety and providing optimal routes.

Services are also envisioned for alleviating the stress of transportation. Modern people are constantly interacting with information spaces, trying to more effectively spend their valuable time each day. They are constantly searching through a sea of information, pressed to make decisions on things like time slots available at the hospital in the morning, crowding and menus of restaurants for lunch, the optimal cafe for a meeting, the shortest route from the station, and so on. This deluge of information consumes a lot of time, and is a psychological burden. In MaaS enabled by IOWN, the service itself will understand the needs of users—based on innovative information gathering and analysis capabilities—and thereby promote optimal situational decisions and information selection. People will be liberated from time-wasting information decisions, and gain fulfillment by spending valuable time more naturally.

③ Navigation acting on the heart

MaaS will enable more efficient transportation. However, the feeling of happiness people experience when traveling does not derive from optimization. When traveling even a short distance by car or train, there are emotionally moving moments, as one experiences the changing scenery with all five senses, and encounters the unexpected. The best approach to optimization changes if we take into account these charms of travel. Sometimes it's good to drop in somewhere on the way, or to take a route with no destination.

"There's a place near here with beautiful scenery where you can spend some peaceful time." "There's a sale on a brand you like. It starts today. That one-piece dress you've had your eye on is half-price." What if there is a service that talks naturally to the user by noticing the user's mental state? Traveling will become something that touches people emotionally. MaaS based on IOWN may investigate the physical and mental background of a person based on bio-information and behavioral history, and support new values and discoveries.

④ Advice from a future perspective

As societal aging progresses, attention is focusing on "healthy aging"—i.e., the ability of elderly people to live their daily lives independently, without long-term care. There is also the worrying situation that one in four people has metabolic syndrome[3] and is at risk of unhealthy aging. How people move around in their daily life is an important issue relating to health. When deciding whether to ride the bus or walk, vacillation might be reduced if one could see, before one's eyes, one's future self inferred from one's current state of health.

Technology for sensing and accumulating bio-information and lifestyle information, without burdening the user, has advanced dramatically in the last 10 years or so. By linking these advanced technologies, MaaS can gain a window onto the daily life of each individual, offering travel advice with a future-oriented perspective. The system will naturally protect the health of each person, and that too is a goal of IOWN.

[1] "Future Investment Strategy 2018 — Reform for "Society 5.0" and a "Data-driven Society" (In Japanese)
https://www.kantei.go.jp/jp/singi/keizaisaisei/pdf/miraitousi2018_zentai.pdf
[2] 4th Public-Private Conference toward Air Mobility Revolution, December 20, 2018 (In Japanese)
https://www.meti.go.jp/shingikai/mono_info_service/air_mobility/004.html
[3] Fiscal 2015, Ministry of Health, Labour and Welfare "Data on Specific Medical Checkups and Specific Health Guidance" (In Japanese)
https://www.mhlw.go.jp/bunya/shakaihosho/iryouseido01/info02a-2.html

Medicine + IOWN

Utilizing medical information to benefit individuals and society

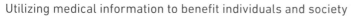

More friendly medicine for both patients and doctors

Due to progress in ICT, there have been major changes in the state of medicine in the last few decades. Medical institutions have seen increasing digitization of medical information such as patient records and test data, and imaging technologies such as CT and MRI based on sophisticated information processing have reached the point of supporting judgments by doctors. Doctors have also begun to refer to databases containing a vast amount of disease information and journal papers from the past, as well as diagnosis support information developed from AI analysis of medical images.

For a person receiving medical care, on the other hand, there has been a revolution in information on one's own body. For example, genome analysis technology and services are increasingly well-developed, and if desired, anyone can receive a genetic diagnosis. It has become possible to diagnose things like risk of disease onset and the effectiveness of medication. Furthermore, wearable devices for monitoring bio-information have become lighter and more compact, and this is making it easier to acquire a constant stream of daily bio-information representing physical activity, heart rate, blood pressure, body temperature, and so forth. It has reached the point where individuals themselves can directly confront the risks and signs of illness, and be aware of their current and future health. Going forward, these approaches will diversify in step with medical research and systems. For example, we may see measures to prevent illness before it occurs, or development of drugs that are optimal for individuals.

On the other hand, diagnostic information and bio-information are the supreme example of private information. Initiatives are underway to actively use these sources of information, but medical information systems are in a transitional stage, and many people have concerns about the handling of information. Going forward, what sort of technology will be used to protect personal information, and help promote the health of individuals and society? IOWN technology will be essential for that future.

4 Use Cases

① Safe and natural telemedicine

Many people are familiar with the da Vinci surgery support robot. This system was developed in the U.S. in the late 1980s to perform telesurgery on people wounded on the battlefield. However, in 2000 it was authorized by the Food and Drug Administration (U.S.) as a laparoscopic surgery system for non-military patients. The robot has been widely disseminated, with more than 4500 units already in service world-wide, and more than 300 units delivered to medical institutions in Japan. With this system, a number of holes about 1-2 centimeters in size are opened in the body, and then tasks like 3D photography of the affected area, removal of tissue, and suturing are performed using a robot arm inserted from those holes. Surgery is done by a doctor at a remote location, using local lever controllers while viewing transmitted images. The system has advantages like small minimally-invasive incisions, and reduced pain, bleeding, and infection, and coverage by insurance is expanding, even in Japan.

In Japan, where depopulation and aging are progressing, primarily in rural areas, the need for this sort of telemedicine is expected to rise in the future, and efforts are being made to reform institutional frameworks [1]. If current communications technology is used for telediagnosis and surgery, we will inevitably run up against limits on the reproducibility of 3D high-definition video, tactile feedback, and communications stability that are essential for diagnosis and treatment by doctors. In the conventional system, video data captured by a camera is converted to electrical signals, compressed, divided into small chunks (IP packets) and transmitted. The signals are then recombined, decoded, and displayed on a monitor on the doctor side. This sort of transmission process, essential in current communications technology, gives rise to data degradation and delay. If the greater capacity and lower latency of photonics technology can be applied to this process, it will enable high-quality communication without data degradation.

The computational and communications capacity of IOWN will bring out to the fullest the capabilities of ultra-high-definition miniature 3D cameras, ul-tra-delicate haptics (technology for skin sensation feedback), and advanced robot-ics—all areas where development is steadily moving forward. For both patients and doctors, the result will be more natural, friendly surgery.

② Pre-learning and navigation systems for surgery support

Under the tension of surgical settings, a small mistake, such as inadequate communication or mistaken confirmation, may result in a severe incident. More realistic pre-surgery simulation and navigation during surgery are promising ways to ensure surgery is properly brought to a successful conclusion. Various advanced technologies are also currently accelerating their evolution and be-coming increasingly sophisticated—e.g., big data sharing and AI-based analysis technology for medical images, 3D output technology, and VR technology.

In navigation during surgery, for example, current efforts include 3D assem-bly of images obtained before or during surgery, display of images overlaid with simulation data or side-by-side with actual echo images during surgery, provid-ing surgery assistance synchronized with real time, and assisting surgery by using actual-size 3D models of organs output in 3D from images obtained before surgery.

The history of employing these techniques in actual surgical settings is ac-cumulated as big data together with surgical cases, and used to update medical knowledge. It is also used in pre-conferences for the subsequent surgery, pre-learning, and educational for future doctors, and so forth. There are high expectations for IOWN, due to its reliable communication and computation infra-structure, as infrastructure for these support systems.

③ Secure, worry-free medical information systems

Medical information in the medical history and DNA information of a patient is the most sensitive type of personal data, and information leaks are never per-missible. Furthermore, if there is trouble with communication lines in a telemed-icine setting, there is a possibility of direct harm to a patient's body. Medical in-formation requires a communication environment and information management system with both the utmost security and low latency. The Ministry of Health,

Labour and Welfare recommends that medical information be communicated by VPN (private closed network), but even that approach depends on specific operational procedures, and is not always safe.

With IOWN, specific wavelengths can be dedicated to sending/receiving medical information thanks to the characteristics of photonics. These dedicated lines can provide an environment for worry-free operation of surgical robots in the context of telemedicine. This new, more reliable foundation will enable smoother gathering, accumulation, analysis, and use of medical information, as well as real-time telemedicine including telediagnosis and telesurgery.

④ From medical information to a healthy future

An environment for handling medical information is currently being rapidly developed [2]. Collecting, analyzing, and sharing medical examination data, diagnostic images, genome analysis/trial data, and other information will drive medical progress, and eventually help the society at large to promote health, prevent illness, and reduce medical expenses. On the other hand, this medical data will be used for purposes like order-made medicine and development of molecular target drugs, and in this way, it should also provide individuals with highly-effective, optimized treatment with fewer problems like side effects.

IOWN will contribute to a secure medical infrastructure to achieve these objectives, and the Digital Twin Computing aspect of the IOWN infrastructure will be useful for future prediction. IOWN will assist personal prevention by, for example, predicting a person's near-term state of health based on genome, lifestyle, eating habits, and preference data, and predicting risks of future illnesses while they are still in the pre-illness state.

[1] Ministry of Health, Labour and Welfare, *Guidelines for Proper Implementation of Online Medical Care*, March 2018 (In Japanese)
https://www.mhlw.go.jp/file/05-Shingikai-10801000-Iseikyoku-Soumuka/0000201789.pdf
[2] For example, the Act on Anonymously Processed Medical Information to Contribute to Medical Research and Development which came into force in May 2018 specifies that handling of anonymously processed medical information to contribute to medical research and development is a special case of the Act on the Protection of Personal Information (amended Personal Information Protection Act). (See the following link in Japanese)
https://www8.cao.go.jp/iryou/gaiyou/pdf/seidonogaiyou.pdf

Finance + IOWN

A safe and secure financial society through optical transmission and encryption

On the eve of the cashless economy

Finance might be called the "lifeblood" of the economy. It is the foundation of human economic and social activities, and money circulates based on safety, security, and trust. In the world we have been familiar with for a long time, money (currency) is issued by a central bank, and physical bills and coins are used to purchase things, with financial institutions licensed by the government acting as intermediaries. This system is undergoing major changes in the 21st century. The keyword is "fintech."

As the word suggests, "fintech" is technology relating to finance. The concept is broad, and includes virtual currencies, robo-advisors for investment and asset management, smart payment (cashless and mobile), as well as remittances and insurance. Fintech is driven by dissemination of the Internet and ecommerce, emergence of new systems such as block chain (the foundational technology for virtual currencies), and the dissemination of smartphones.

Cashless payment—one aspect of fintech—is being strongly promoted in Japan by the national government. The government's target is to raise the cashless payment rate of 18.4% in 2015 to 40% in 2025 [1]. Japan has a stronger attachment to cash than countries like the U.S., which has a high rate of credit card use and a tradition of payment by check, and South Korea and China where there has been a rapid shift to cashless in recent years. However, even Japan has provided some support—e.g., reward points for increases in consumption tax—and is finally making major strides toward cashless.

For individual users, technologies like cashless payment (mobile payment) raise worries about theft when money is electronically exchanged. In fact, there are an endless number of ways to deceitfully steal money in electronic transactions. Techniques used include unauthorized use of accounts via cracking (unauthorized access to systems connected to a computer network) or identity fraud, theft of virtual currencies, and fraudulent ICOs (Initial Coin Offerings, i.e., public offerings of a new virtual currency). Many systems being developed today do not adequately ensure safety, security, and reliability.

Secure optical transmission and encryption technologies realized by IOWN are essential for the truly safe and secure financial society of the future.

3 Use Cases

① Secure transactions and asset circulation using encryption

One aim of the Smart World realized by IOWN will be an all-photonics network and, eventually, end-to-end optical transmission.

The distinguishing characteristics of optical transmission are high speed, low latency, and security (safety). If transmission is carried out using end-to-end optical transmission—assigning wavelengths to each service—we can minimize the rate of electronic information processing, and greatly reduce the risk of cracking during transmission.

Quantum-resistant encryption is one encryption technology we will use. Today, there are worries that the encryption algorithms securing conventional ICT systems will be broken in an instant by quantum computers. Quantum-resistant encryption will be used because it is resistant to such attack.

There are fears that the first target will be financial transactions when quantum computers finally reach the practical level. R&D is currently underway at NTT on quantum-resistant encryption that can withstand such attacks.

② Automated asset management based on values

Today, we are seeing broader adoption of AI-based investment support services called "robo advisors," and use of AI by Internet securities companies (e.g., proposal of optimal asset allocation, automation of investment management, and AI support for improving user trading techniques). Typical variables of robo-advisors include risk tolerance for investments, but we envision support systems for automated asset management that make decisions reflecting the values specific to each person. Asset management in line with each person's desired life plan will be achieved, without any need for attention by the person, while always reflecting the person's values, thinking, and behavior patterns. The desired life plan will be formulated by making timely proposals for asset management, and encouraging behavior modifications based on the values of each individual. These support systems for automated asset management will be one application in the Smart World shaped by IOWN.

③ Payment without cash registers or ticket gates

Amazon Go stores with no cash registers are being rolled out in the U.S. by Amazon.com. The stores are creating a lot of buzz as a potential model for retail in the near future. These are reports that these stores will be expanded to 3,000 locations by 2021. Similar register-less stores are already operating at many locations in China. In September 2019, NTT Data began offering the retail industry in Japan services to support register-less digital stores. At these stores, people can simply pick up items and take them home without paying at the register, provided they obtain authorization when entering via a QR code with designated means of payment [2].

If it becomes possible to do instant, automatic payment whenever there is an exchange of goods or services, with no need to pass through a register or ticket gate, it will save time for users, who will no longer have to wait for the register. On the store side, it will reduce labor costs, and save labor through things like automated product management. It will also eliminate the severe problem of product shoplifting that is always associated with retail.

We believe it will be possible to achieve a world with no registers, ticket gates, or other gates, if we can establish a system of natural payments in real time—through a combination of authentication technology, proper handling/use of personal information, and natural interfaces.

[1] Ministry of Economy, Trade and Industry, *Cashless Vision*, April 2018 (In Japanese)
https://www.meti.go.jp/press/2018/04/20180411001/20180411001-1.pdf
[2] NTT Data, "News Release," September 2, 2019 (In Japanese) https://www.nttdata.com/jp/ja/news/release/2019/090200/

Experience/Learning + IOWN
Optical communication to expand the horizons of life

Novel emotionally-moving experiences through communication

In the last decade or so, ICT has penetrated deeply into people's lives, and our perceptual and emotional experiences, and how we experience physical activities, are different from what they were before. The boundary between the real and virtual has already become blurred among modern people.

For example, you might prepare a new dish while watching a recipe movie app and post the results to social media; or engage a teacher at a volunteer cram school, popular in the Internet ratings, to teach you how to solve math problems over the Internet. Communication across time and space, via connections provided by ICT, stimulates curiosity and motivation in each person's heart, and beckons people to try new experiences.

The year 2016 was designated by some as the first year of the VR (Virtual Reality) era, and firms throughout the world have embarked on VR-related businesses for ordinary users. VR technologies—such as surround sound and head-mounted displays providing a view of spatial video all around oneself—boost emotional response by enhancing presence and immersion. New initiatives have been launched in many fields, including esports and games, simulated experience tours (as with Google Earth VR), simulated experiences of architecture and design, learning and skill instruction, high-presence live broadcasting, and multi-modal movie viewing.

Research is also moving forward on AR (Augmented Reality) technology for overlaying elements of cyberspace onto real space, and mixed reality where those elements are integrated within a virtual space.

For a long time at NTT, we have been engaged in research to intensify VR and make it more comfortable. We have also conducted research on immersive telepresence for experiencing live events at remote locations with a sense of ultra-high presence. The advent of IOWN will drive these technologies further in the future, and realize still-unknown realms of emotionally-moving experience.

3 Use Cases

① The ultimate form of "you are there" experience stimulates the five senses + something extra

At NTT, we have conducted R&D since 2015 on the immersive telepresence technology "*Kirari!*" for real-time delivery of the total experience of a stadium or live space. As part of that, we have developed technology for highly-efficient compression of media data—representing players/performers, background images, and audio—as well as technology for intricate synchronization, reconstruction, and composition. High-presence media, enabling many people to experience sporting and other events, may become one of our future legacies.

The communication and computation environment of IOWN will further evolve these technologies, and open up new, unknown domains. In ordinary life, people's experiences and emotions are produced by interaction between their environment, body, and inner world. Environmental elements acting on the body go beyond the scenes and sounds captured by vision and hearing, and include things like air temperature, humidity, air pressure, electrostatic voltage, UV intensity, oxygen and carbon dioxide concentration, gravity, magnetic fields, wind, and sound pressure. The delicate sensory organs and systems of the human being shape experience as a whole by unconsciously capturing some parameters, and consciously capturing others.

The optical technology of IOWN aims, as far as possible, to transmit the interaction between a person and the environment as is. By capturing changes that are stripped off with conventional digital technology, and transmitting them in an appropriate, multi-modal fashion, the experiencer can receive them naturally with his or her whole body. A full-body experience where the experiencer feels the unchanged, analog feeling of the air felt by a performer at a remote location provides a sense of presence that people have never before experienced.

② The ultimate in training with Digital Twin Computing

How far can a person truly re-experience the experience of another person? With the aid of IOWN, we will continue to pursue this possibility. At NTT, we are currently engaged in research to analyze video, bodily movements, and bio-information during sporting competitions or training, and this research will be put to work improving the performance of athletes. Based on the results, we should be able to create revolutionary training systems employing IOWN that integrate the cyber and physical side.

For example, by using Digital Twin Computing technology, it may become possible to recreate the performance of an athlete, of similar to oneself, and communicate the secrets of elite technique using something like robotics technology.

However, for this to be genuinely useful in actual competition, it is not enough to simply re-experience the form, sights, and sounds. There is a need for deeper knowledge of the human being itself. We want to hone in on the essence of the human by analyzing a wide range of information, and incorporating results in fields such as brain science and cognitive psychology that have made major progress in the last few decades. We want new experience research involving IOWN to be based on scientific understanding attained by investigating the question "what is a human being?"

③ Augmentation of experience will change how we communicate

Future VR technology will go beyond vision and hearing, and will enable sending and receiving of more comprehensive information in real time. This will be achieved by scanning a more multi-faceted range of sensory information including: taste and olfaction, vestibular sensations indicating the inclination of the body, and somatic sensations such as cutaneous sensation, deep sensation, and visceral sensation. Furthermore, evolution of the human machine interface (HMI) may enable realistic simulated experience of the sensations of another person. Such experience will promote mutual understanding, create new opportunities for communication and collaboration, and offer opportunities for new experiences.

In recent years, there have been news reports of patients—with difficulties going out due to intractable disease—attending weddings of relatives, or achieving a long-cherished desire to climb mountains by using an avatar robot. In the future, IOWN will further free us from physical sensations, and amplify our experiences.

We might call this a technology which extends the way people live.

Public Service + IOWN

Protecting society from disasters and crime

Future prediction technology and Smart Cities

All over the world, there is a shared desire to protect people from natural disasters, accidents, and crime, and create a safe society where everyone can live with peace-of-mind. If disasters and crime can be accurately predicted, it will be possible to minimize damage. Through IOWN, we will realize high-definition sensing, utilization of big data, and increased sophistication of network technology, and this will help us meet the challenge of difficult problems.

Especially today, when natural disasters are a frequent occurrence, there is a pressing need to establish information networks within local governments to predict disasters—as accurately as possible—and provide prompt notification to residents. The "Future Investment Strategy 2018" promoted by the Japanese government calls for building Smart Cities as one flagship project in a high-priority field.

If these networks are realized, they will enable uniform broadcasting of information, as well as close support that is tailored to individuals. When a disaster or incident actually occurs, it will be possible to keep damage from spreading by providing accurate guidance suited to the situation.

Networking will also bring major changes to the methods of administrative work. Individual resident information will be shared within the government as appropriate, and complicated application procedures, previously carried out at the counters of city offices will no longer be necessary.

3 Use Cases

① Predicting disasters with optical technology, and using for disaster prevention/mitigation

Theoretically, it is difficult to predict the occurrence of disasters like earthquakes, volcanoes, and torrential rain based only on their underlying mechanisms. A more realistic approach is to carry out simulations—by observing current conditions in detail, and comparing with past data—and then probabilistically predict where, when and what sort of disasters will occur in the future.

High-precision data is necessary to predict the future with high probability, and observation methods using optical technology are currently attracting a lot of attention.

Optical lattice clocks [1] are extremely accurate clocks which deviate only 1 second in 30 billion years, and can measure even the slight deviations in time due to the earth's gravity. Using such clocks, it is possible to conduct relativistic surveying of altitude differences with error on the order of centimeters [2]. In addition, optical fibers can detect distortion on the order of a few microns. With these techniques, it is possible to capture very slight indications of volcano movement or landslides.

If various types of observation data are linked by a network, it will be possible to capture phenomena over a wide range, and further improve prediction precision. It has been proposed to build a network of optical lattice clocks, with clocks set up at various locations, and if this plan is realized, it will be possible to precisely monitor crustal movements (e.g., volcanic activity and plate movements resulting in changes in elevation) that occur on a time scale from a few hours to a few years.

There are also high expectations for establishment of a GNSS (Global Navigation Satellite System) and an ultra-high-precision system to measure elevation differences to complement GNSS. The distinguishing features of IOWN—high capacity, low-latency transmission technology, and large-scale computational processing capability—will make a major contribution to achieving these systems.

② Smart Cities for proactively preventing incidents and accidents

The mass shooting which occurred in Las Vegas (Nevada, U.S.) in 2017 shocked the world. Crime prevention awareness is high in Las Vegas, and NTT is current collaborating with the city on a Smart City initiative. Crime prevention is a major theme of this initiative.

In this project, the entire city is being monitored with numerous cameras to proactively prevent incidents and accidents. Information (e.g., strange noises, suspicious persons, people gathering, or a vehicle driving the wrong way) is integrated in real time, and situations that may occur in the future are predicted with a specific probability. If, based on these predictions, government agencies like the city or the police deem it necessary, the police or fire department is dispatched, and measures are taken to proactively prevent an incident or accident. If an incident or accident has already happened, the system provides support in real time for announcements to the public, and deployment of police and fire trucks.

ICT resources must be allocated quickly and efficiently when handling big data, and therefore the system incorporates Cognitive Foundation.

Naturally, handling of personal information must be considered while also taking into account the importance of crime prevention. In Las Vegas, the city takes responsibility and has a system for managing data. We propose a system where service providers do not accumulate data, and data is properly managed by local governments.

③ Seamless integration of information to eliminate filling out forms

In recent years, local governments have been increasingly switching to smart administration—in which work is automated using AI and robotics—due to concerns about future labor shortages. Various methods are being adopted to improve work efficiency such as IVR systems for automatically answering telephones, chat and social media response, and AI-based chatbots.

If all the events in the life of an individual—birth, school advancement, and examinations at medical institutions—are recognized in the administrative system, then there is no need to file forms in the first place.

However, this convenience has danger as its flipside. This sort of system can only come into being through the linkage of resident information, financial information, and administrative systems, and therefore thorough measures must be taken to manage personal information. An information processing infrastructure will be needed to realize secure data distribution across industries.

[1] A next-generation atomic clock conceived in 2001 by Hidetoshi Katori, at that time an Associate Professor, School of Engineering, The University of Tokyo. First, atoms laser-cooled with a periodic energy wave (optical lattice) are captured. This lattice is created by interfering laser light with a special wavelength called the "magic wavelength." The setup ensures there is no interaction between atoms. Next, a laser light is shined on these atoms, and the oscillation frequency (resonance frequency) of the absorbed light is precisely measured. The length of 1 second is determined from this frequency. The optical lattice as a whole can capture multiple atoms, and thus time can be determined at high precision in a short time by measuring the resonance frequency of those atoms all at once and taking the average.

[2] According to Einstein's general theory of relativity, if two clocks placed at different heights are compared, the lower clock is more greatly affected by the earth's gravity, and its time ticks off more slowly. Differences in altitude can be determined by placing optical lattice clocks at two different locations, and measuring the difference in how they mark time.

Energy + IOWN

Solving energy problems with technology for optical power supply, energy production, and energy storage

Revolutionizing energy systems

The development of human civilization in every era has involved acquisition of energy, but energy problems are especially serious for modern society, which depends on electrical power for so many things, including food, clothing, shelter, essential utilities, industrial production, and services. With increasing global population and rapid growth of developing countries, global energy consumption will continue to increase, and estimates predict that 1990 levels will be roughly doubled in 2030. Reduction of the greenhouse gases that cause global warming is also a pressing issue, and most emissions derive from combustion of fossil fuels (e.g., oil, coal, natural gas) to produce energy.

The Paris Agreement, signed in 2015 at the United Nations Framework Convention on Climate Change Conference of the Parties (COP), establishes as goals: keeping the increase in global average temperature well below 2°C above pre-industrial levels, pursuing efforts to limit the increase to 1.5 °C, peaking greenhouse gas emissions as soon as possible, and achieving a balance of greenhouse gas sources by sinks in the second half of the 21st century. All participating countries and regions, including developing countries, are required to establish targets for greenhouse gas reduction/control from 2020 and onward, and must work to draft and submit a long-term strategy for low-emissions development. In response, Japan set a medium-term target of reducing greenhouse gas emissions by 26% compared to fiscal 2013 levels by 2030. The Fifth Basic Energy Plan announced in 2018 calls for an 80% reduction of greenhouse gas emissions in 2050.

To achieve these ambitious goals, it is essential to make a large-scale shift away from energy consumption dependent on fossil fuels, and this will require innovation in all the technologies and systems of the energy mix, including power generation, storage, transmission, and distribution. IOWN technology will be an important key to solving these difficult problems and promoting innovation.

2 Use Cases

① Optimization of distribution to support energy production, conservation, and storage

At present, fossil fuels such as oil and coal account of more than 80% of Japan's electrical power generation. Renewable (natural) energy, such as solar, wind, and biomass, has rapidly grown due to adoption of feed-in-tariffs (FIT) [1] in 2012, but these sources still account for only 16% of the total (including hydropower in 2017). Efforts are being made to increase that rate to 22-24% in the 2030 energy mix, and to even higher levels as we move toward 2050. However, there are problems too.

Power generation and distribution systems for establishing renewable energy as a stable power supply are insufficient where they are needed, and sharing with existing transmission systems is also difficult. Progress must be made in developing new technology for storage and output control of renewable energy because it tends to vary with the seasons and weather. Systems must also be developed for local production and consumption. Huge costs will be incurred developing such infrastructure.

NTT is tackling these difficult problems through innovation in photonics technology. For example, we envision using optic fibers as transmission lines. Today's optic fibers, however, were developed for communication, and cannot withstand power input and output at the power transmission level. To address this, NTT is currently collaborating with other firms and universities, and conducting R&D on multi-core optic fibers with improved internal structure of the fiber core (the section through which light passes). This research on optic fibers for energy transmission also involves examining new materials. R&D on wireless power delivery is progressing all over the world, and if it becomes possible to easily deliver power via optic fiber, this will enable flexible supply of energy to sensors or robots located virtually anywhere. It will also allow more effective use of optic fiber networks, and help optimize distribution of electric power.

If improved computational capabilities of the information processing infra-structure enable more precise electricity storage and on-demand output control, energy distribution will be more efficient. By supporting diversification of energy production sources and optimization of energy distribution, we will create a smarter relationship between people and energy.

② "Artificial photosynthesis" — the ultimate in clean energy

Most of the earth's energy derives from the sun. Solar energy produces the cycles of the atmosphere and hydrosphere, and drives the photosynthesis of plants and the activity of life. Traced back to their source, fossil fuels are ancient plants and animals, made of carbon compounds produced by the sun's energy and accumulated over hundreds of millions of years.

Effective utilization of sunlight has the potential of solving our energy problems. One approach is solar power generation, but the amount of power produced is unstable, and varies depending on the amount of sunlight. Solar also has the disadvantage of requiring costly maintenance. So what other methods are available?

Artificial photosynthesis is one possibility, currently garnering attention in cutting-edge research all over the world. If the photosynthesis of plants can be artificially recreated, it will enable production of clean energy from water, sunlight, and carbon dioxide in the atmosphere. It will also reduce the carbon dioxide that causes global warming. In addition, the synthesis process will enable production of fuels like hydrogen and methane that are likely to see use as new energy sources.

NTT is currently engaged in R&D on artificial photosynthesis technology, leveraging the semiconductor growth and catalyst technologies we have cultivated in previous R&D on optical communication and batteries. This technology will be able to produce hydrogen, methane, and other fuels from water and carbon dioxide by receiving light from the sun, just like the photosynthesis of plants.

Artificial photosynthesis is a technology for realizing a "carbon cycle society" which cycles carbon through the following stages: carbon dioxide carbon compounds energy carbon dioxide... If we can realize such a carbon cycle society, powered by the sunlight that strikes the earth, the earth will return very close to its original natural state. In the world of IOWN, our goal is to promote artificial photosynthesis and realize an unprecedented clean energy society.

[1] System for purchasing electricity generated by renewable energy at a fixed rate for a fixed period of time.

Part 2

Technology toward IOWN

Chapter 5

11 Technologies and 20 NTT Case Studies

NTT Research and Development Planning Department

01 Artificial Intelligence
Developing AI that has *Generosity* and *Sincerity*

There are high expectations for AI throughout a broad range of industrial sectors. Research efforts are flourishing around the world, and the market is slated to escalate at a high 51% per year from 2018 to 2025 [1]. Research being conducted in each country is delivering more technology of practical use, and there will indeed be an increase in the number of heated debates regarding the future direction of AI development and new types of applications.

There have been a slew of research papers published in the field of AI in recent years, mainly from the United States and China. Popular topics include *streamlining training data*, *white-box AI*, and *faster speeds with purpose-built architecture*, and today research institutions around the world are jostling to be at the top of their field.

3 topics

Current development of AI using machine learning or deep learning generally requires vast amounts of data to be made available. An example of this is the medical or financial fields, where AI applications are expanding—and becoming fragmented—at an ever increasing rate. In such cases, data needs to be collected, stored and analyzed for each specific field of application, resulting in an explosive growth in associated costs. One way of addressing this that is gaining attention is *streamlining training data*. Instead of using vast amounts of data, being able to learn efficiently using as little data as possible can help slash data collection and other associated costs significantly.

One approach for efficient learning with less data that is gaining traction is *transfer learning*, where knowledge gained to address one issue is applied to a different issue. An example of this is when breast cancer is identified through mammography images using deep learning. By applying transfer learning using a neural network with knowledge acquired in advance from ordinary images on ImageNet, identification accuracy can be increased even with a limited number of mammography images. [2]

Deep learning also usually takes an automated approach to learning from vast quantities of data, to derive weighting coefficients for a neural network comprising a large number of layers to bring about results. These processes are too complex for humans to grasp, and as such the *black box* approach of AI processes is increasingly being viewed as a problem. Without any understanding of the internal processes being run by the AI, it becomes difficult to trust the results obtained. To overcome this, there is growing demand for *white-box AI* to gain a better understanding of the delivered results. Research into such methods is gaining traction, with an example including the *Explainable AI* project launched in 2016 by the Defense Advanced Research Projects Agency (DARPA) to study ways to achieve artificial intelligence with more transparency.

High-performance hardware is also required to achieve AI of such an advanced level. Yet there are hurdles that need to be overcome, as CPUs and GPUs used for ordinary applications lack the performance required to run vast numbers of complex calculations simultaneously, and immense computing resources are needed. For that reason, development of purpose-built architecture and frameworks designed specifically for these AI calculations will become vital.

Shifting from *Seeing*, *Talking* and *Listening* AI, to a more advanced level of *Thinking* AI

At NTT we are keeping abreast of trends into these fields of research, and working to advance AI to the next level in line with our policy of delivering the benefits of ever more advanced and complex cutting-edge technologies to society in a natural and seamless manner. To enable AI to exist naturally alongside people from all walks of life, AI must not only be able to approach explicit problems in a uniform manner, but also needs to know the reasons behind the behavior each person takes, understand the sense of values held by each individual, and independently support the actions that people take from an extensive range of perspectives. To achieve this, AI must make the transition from simply *Seeing*, *Talking* and *Listening* to combining these as a *Thinking* AI capable of more advanced, logical and analytical thought processes.

To successfully develop a *Thinking* AI capable of processing the values of individual people as well as adhering to its own values, the AI must first be equipped with the very fundamental abilities that humans possess—*seeing*, *talking* and *listening*. An example of the *Seeing* ability is *Angle-free Object Information Retrieval* that NTT is developing, which enables high-precision recognition of objects changing shape, from very few reference images. Advances are being made to *Listening* and *Talking* abilities with case studies and research into speech recognition and spoken dialogue technologies through ongoing discussions with Professor Noriko Arai of the National Institute of Informatics and the pioneer of android research, Professor Hiroshi Ishiguro of Osaka University.[3]

Humans also have the natural ability of crossmodal processing of information consolidated, supplemented and spanning numerous media sources on an ordinary basis. An example of this is that even with their eyes closed, humans are able to create an image of the scene transpiring around them just by listening to the sound. NTT is actually addressing this very topic, as it is focusing on developing technologies capable of recognizing sounds and generating images of that particular scene—deducing images from sounds. Research into crossmodal information processing capabilities like this represent the first step toward AI that is capable of *Thinking*: where AI can respond to and recognize aspects of the outside world through different sources of media like images and sound, and use that information to acquire a fresh new *perspective*.

AI research with a focus on *Tolerance* and *Sincerity*

By understanding the diverse range of values that humans possess based on these fundamental capabilities, we are seeking to expand on the fields that AI can play a role in by developing our own *Thinking* AI. Even with today's technology, AI is able to translate simple text documents or handle customer support inquiries via a chat interface. Yet these cases are only delivering the optimum responses corresponding to the information provided. Nonetheless if, for example, AI is able to provide new suggestions or raise problems based on the experience of the AI itself, they should be able to communicate in a way that broadens the way humans think while expanding on the actions available to them, instead of simply answering questions.

And while communicating with AI, if the AI is able to deduce the sense of values of the person it is talking with and then respond with its own values, the result will be a richer, more creative conversation. These abilities are significant, particularly when adopting AI throughout a broad range of fields such as counseling or aged care facilities. Continuing to chat with AI as a form of communication will eventually help to build up *trust* in AI as a valuable conversation partner for humans. Instead of simply coming up with the optimum solution for any particular situation, the resulting AI will be able to focus on long-term goals and propose various actions to increase the range of options available to humans, and even help us think on a much deeper level.

In examples such as route navigation based on forecasts of a person's movements or state of mobility, an understanding of the diverse range of values that individuals have is important. If AI is able to process information based on various values not limited to casual chats, conversations or route navigation, it may be able to open up the options available to humans even when facing more complex problems.

The goal of developing such an AI built on a diverse range of values stems from the concepts of both *Tolerance*—being able to accept myriad ways of thinking in a flexible manner, and *Sincerity*—earning a deep level of trust from humans by responding without any inconsistencies or breakdowns. Until now, AI has taken an automated approach to learning data generated with each passing moment, and responded based on the biases included within that data. The resulting presence of AI is that it conveyed both the positive and negative aspects of human society. To bring AI even closer to humans, it will be important for development to include the *Tolerance* to be open to a broad range of ideas, and the *Sincerity* to respond consistently while behaving in flexible manner that is accepting of diversity. And to ensure that AI is capable of even deeper *Thinking* as a way of assisting humans' thought processes, we need to focus on developing AI that is built with *Tolerance* and *Sincerity* while further advancing research into AI.

[1] Allied Market Research, Global Artificial Intelligence (AI) Market, 2018-2025
[2] Examples such as papers published in *Physics in Medicine and Biology* Vol. 62 (Samala R. K. et al., "Multi-task transfer learning deep convolutional neural network: application to computer-aided diagnosis of breast cancer on mammograms")

[3] NTT took part by covering the English field of the "Can a Robot Get into the University of Tokyo?" project run by Professor Noriko Arai of the National Institute of Informatics, and provided the required language and knowledge processing technologies. NTT also co-developed the android with Professor Ishiguro, with his likeness "Geminoid HI-4" including NTT's casual conversation, speech recognition and clear sound pick-up technologies in noisy environments to successfully have natural conversations with humans. https://www.sxsw.com/news/2017/a-casual-afternoon-conversation-with-droids/

Case Study 1

Speech recognition and casual conversation technologies
Aiming to have more natural conversations

We are focusing our efforts on R&D into various technologies related to communications, including speech recognition systems capable of accurately understanding the intentions of the speaker, and automated response technologies utilized by chat bots. *totto* **(Fig. 1-1)** is an android that was designed based on the actress/personality Tetsuko Kuroyanagi. In addition to being equipped with conversation skills based on vast amounts of data, it has been developed with many of Ms. Kuroyanagi's character traits that it learned from shows that have been broadcast. By utilizing high-precision speech recognition technology and highly reproducible voice synthesis technology, as well

Fig.1-1: *totto*
©2017 totto production committee

as conversation processing technology to simulate natural behavior complete with character traits, users enjoy communicating with the android as if they were actually speaking with Ms. Kuroyanagi in person. Combining these with automated motion generation technology that matches the conversation means that robot control technology can create even more natural conversations.

Applying extensive experience and advanced deep learning
Today, high-precision speech recognition technology is being utilized for a multitude of applications, and NTT has a long track record with speech recognition R&D spanning half a century. When we started out, the technology was only capable of

recognizing clearly spoken written texts like newspapers, and the vocabulary it could recognize was also severely limited. Yet we were the first in Japan to utilize technology called WFST (Weighted Finite State Transducer), which made it possible to recognize the best word out of a database comprising 10 million words—a 100-fold increase over existing technology at the time. And by applying deep learning technology that is gaining attention in recent years, NTT scored first place in an international challenge (CHiME3 2015) assessing the accuracy of speech recognition systems on mobile devices used in noisy open areas.

Built on the fruits of years of such research efforts, we developed the *VoiceRex®* speech recognition engine for use in a diverse range of applications, which we are now supplying to group companies. An example of one application is in the call center business. Eliminating background noise and other information from conversations recorded with the permission of customers to identify data patterns efficiently helps to provide information that can be useful for identifying underlying customer needs or issues with services.

Natural voice synthesis technology that is closer to human speaking is used throughout a broad range of scenarios, like automated response systems at call centers and for entertainment purposes in robots with specific character traits. These days, deep learning is being used for this voice synthesis technology in the same way as speech recognition technology, to create more natural sounding synthesized voices. With past voice synthesis technologies, a large amount of the desired speaker's voice data was required to generate the speaker's synthesized voice. Yet NTT used deep learning and a large quantity of data from another speaker, to successfully create high-quality voice from a small amount of voice data supplied by the speaker.

We are also channeling development efforts to meet demand for multilingual voice synthesis that is growing every year, such as conversational agents for foreign tourists visiting Japan. One of these technologies is *Cross-lingual Speech Synthesis*. To synthesize the voice of a single speaker in multiple languages using conventional voice synthesis technology, voice data of that person speaking multiple languages was required. Yet *Cross-lingual Speech Synthesis* makes it possible to synthesize speech in English as well as Chinese, Korean and various other languages from Japanese voice data supplied by the speaker. This technology allows voice synthesis for multilingual services using the voice of a person who can only speak Japanese. There is heightened anticipation for voice synthesis technology, including synthesizing voices that are more closely associated with human emotions, and we are also focusing our efforts on developing technologies that will be able to meet expectations like these.

Efforts to add character traits and empathy

Research is currently being conducted into conversation processing technology aimed at achieving natural conversations—which is drawing attention due to the increase in popularity of smart speakers and similar devices—as well as research into conversation systems that display certain character traits or that are capable of more complex conversations.

We believe that giving systems like *totto* specific character traits is one way of having users continue to admire and use such systems for a longer period of time. To this end, we are rolling out the *Narikiri AI* project where users take the reins in developing the conversation system of specific mascot character. With this system, users pose questions to their specific mascot character and then reply on behalf of that character, to create conversational data. *Narikiri AI* then uses this data for learning to create a simple conversational system.

Another different project we are researching is a discussion and debate system capable of having debates with users. When users make a comment, the discussion and debate system either supports or objects to the comment. This develops a sense of empathy in the user, which can make conversations more enjoyable or open up new possibilities. Some of the features behind this discussion and debate system was included in *totto* as a function that supports the users' comments. This allows *totto* to empathize with users over a broad range of topics.

We have been involved in communication-related R&D for many years now, covering a multitude of fields such as voice processing and conversation processing. Advances in deep learning technology and the growing areas of applications of AI-related technology will only help to further break down barriers between associated fields, as technologies capable of handling a diverse mix of information—voice, text and images—grow in demand. We will continue pushing our research and development efforts further so that we can provide technology that enhances everyone's lifestyles, by combining various related technologies like we did with the android *totto*.

Case Study 2

Angle-free Object Information Retrieval Technology
Changing the methods used for tourist information or product control

We are increasing research efforts into technology capable of identifying objects when they are viewed at angles other than from directly in front—like you would when searching for images of buildings or products. This is changing the way we can provide tourist information or product control at stores. The *Angle-free Object Information Retrieval* project that NTT has been running since 2015 is an example of this. The technology recognizes objects with a high level of precision regardless of which direction photos of them are taken from, and enables related information to be displayed.

Innovative technology for simplifying recognition of objects

To give AI the ability to *See*, development of technology capable of recognizing and searching for objects in the immediate vicinity is essential. Yet compared to flat items like books, CDs or printed documents, three-dimensional objects can appear significantly different in images depending on the direction that photos are taken. Identifying such objects becomes difficult when photos are taken at different camera angles. In such cases, enhancing recognition accuracy requires preparing many images of each object taken from different angles, adding labels to each image and categorizing them into a database. This workload presents an immense financial burden to service providers, and has been a stumbling block that has prevented the widespread use of services that utilize object recognition.

Angle-free Object Information Retrieval simulates the changes in the three-dimensional view of an object and accurately identifies the relationship between an input image and reference image. This means that the number of images that need to be pre-compiled in a database can be cut down significantly to around 1/10th the quantity. It also uses a statistical method for estimating the priority of image features based on how frequently they appear, which provides a major boost to the search precision.

The technology is being honed even further, where it is currently at a level that is capable of recognizing a wide range of irregularly shaped products like textiles and items in flexible packaging. When non-rigid objects are viewed as an image they go through numerous distortion patterns, which results in a drop in recognition accuracy. Instead of applying the geometric constraints of identical objects calculated through projection geometry to the entire object, the technology applies it to several different sections of the object which enables it to correctly identify the relationship between input and reference images, even if the image of the object is distorted. This method has given AI the capability to recognize and search with a high degree of precision not only regular shaped objects, but irregular ones that appear distorted in almost every way possible.

Info service for inbound tourists and labor-saving at stores

Examples of *Angle-free Object Information Retrieval* in real-world applications are likely to include providing information and guidance at event or tourist venues, or as part of labor-saving efforts for product control or operating cash registers.

There were 31.19 million foreign tourists who visited Japan in 2018, and this number has grown for the seventh year in a row. For foreign visitors, the lack of multilingual displays, and traveling, dining or other activities in unfamiliar areas can be a large cause of stress. Passengers who have just arrived at an airport or train station are faced with numerous difficulties that need to be addressed—most signs are only available in major languages like Japanese and English; scant information on how to get to their next mode of transportation or historic site; or no way to find what ingredients are used in dining options.

To overcome such hurdles, NTT developed the *Kazashite GuidanceTM* service using the technology behind *Angle-free Object Information Retrieval*. This service displays route guidance, detailed tourist information and other tips in the specified language simply by holding a smartphone in front of signboards, buildings, products or other objects. The performance of this service was verified as being of a practical level during a joint demonstration test run between Tokyo International Air Terminal Corporation, Japan Airport Terminal Co., Ltd. and Panasonic Corporation. In this demonstration, passengers using Haneda Airport were able to acquire useful information in their own language by pointing their smartphone at guidance panels, signboards, and menus of certain restaurants.

Kintetsu Railway Co., Ltd. and NIPPON TELEGRAPH AND TELEPHONE WEST COR-PORATION (NTT WEST) also developed a service that provides seamless guidance to users regardless of their age or language. They ran a demonstration experiment for this multi-modal agent AI, combining the functions of *Kazashite GuidanceTM* image rec-ognition AI and *chat bot* conversation AI. The service provides route guidance from stations to destinations or other transportation options in a chat-like format. Users just need to take photos of tourist sites, posters or other images via a dedicated web-site, and the AI recognizes the images to provide the relevant information. The service was available for use by the many passengers using Kintetsu Nara Station, and more than 400 foreign tourists were interviewed about their experience. More than 90% of them responded with answers like "This service is so easy to use" and "I'd like to use it in the future," indicating that the service was well-accepted by users. With examples such as these, *Angle-free Object Information Retrieval* is contributing significantly to information supply services using a broad range of objects in our daily lives as the key.

The latest advances in the technology has made it possible to recognize irregu-larly shaped products like textiles, and items in flexible packaging like snacks and jelly drinks. If the available applications continue to increase in the future and an even broader range of products can be recognized, the technology is likely to be adopted to streamline labor required for cash register operations, and looking further ahead, even used for unstaffed stores. The technology is suitable for product control including sorting work and inventory control, so it will likely have a positive effect on achieving more efficient operations at business sites.

Case Study 3

Spatio-temporal Multidimensional Collective Data Analysis Technology
Identifying human or object movements for more efficient transportation

With the increasing utilization of smartphone and IoT devices these days, there is a great deal of data being measured that is linked in a spatio-temporal manner, including the movement of cars or objects, the behavior of people, and changes in the environment. The collected and stored data is as vast as it is diverse. If AI can be used to learn from this data to identify any hidden information that might be of benefit, it will likely be possible to make forecasts of *time* and *place* in any manner.

Analyzing and utilizing vast quantities of spatio-temporal data

NTT is working on developing *Spatio-temporal Multidimensional Collective Data Analysis Technology* that can forecast *when something might happen, where, and in what way* from spatio-temporal data, by modeling the spatio-temporal relationship of multidimensional data that includes numerous attributes. Collective data analysis is technology used for estimating the spatio-temporal flow of people or traffic in cases where individual objects cannot be recognized, such as when counting the number of people or cars within a spatial mesh, and only the aggregated statistical data is available. As such, this is beneficial in that useful information on a group can be identified while also protecting the information of individuals.

We are applying this technology and working with NTT DOCOMO, INC. to use mobile spatial statistical data as part of studies into the viability of *Near-future Crowd PredictionTM*, which can anticipate the number of people in a specified area at the present time as well as several hours in advance. This technology works by modeling the underlying structure based on chronological data of the population within each mesh, and then learning from pattern changes in this underlying structural model. The result is being able to forecast the population of each 250 to 500 m mesh area, at the present time as well as in several hours time. Research is also being conducted on how to apply this technology to car sharing services or *on-demand buses*, where bus operations can be controlled in a flexible manner based on demand predictions.

This technology is being further refined as part of initiatives aimed resolving traffic problems in urban areas, such as optimizing the flow of traffic and transportation to alleviate congestion and traffic jams.

To reach a solution, we are aiming to roll out the technology for multi-modal MaaS, where numerous forms of traffic and transportation are linked together as a means of transportation with no wastage whatsoever for users. We are focusing our research on the following areas: (1) Import information on the measured flow of people or cars into a simulated environment (data assimilation); (2) Identify as quickly as possible potential guidance alternatives that can alleviate traffic problems in urban areas (optimize control policies); and (3) Present the results as technology capable of guiding groups of people (behavioral change).

Suggest behavior to users for optimizing transportation

(1) *Data assimilation* is required to achieve. This means recreating the flow of people in a simulator. Real-time mobility-related data in urban areas, past historical data, transportation system timetables and the number of people passing through ticket gates are combined with external information like road and station structural details as constant environmental variables, and used as the basis for learning. To recreate the flow of people, the application of data assimilation technology is required to synchronize a diverse range of data and update simulations.

Data assimilation makes it possible to recreate the flow of people by estimating changes in population over time for each transportation route, based on data acquired from pinpoint sources. Modeling also takes into consideration the agent movement time, and the time for each transportation route is estimated that best recreates the data measured from pinpoint sources. Many of these simulations are run efficiently and quickly, to enhance accuracy significantly.

(2) *Optimize control policies* plays a key role for achieving. In this step, details on transportation system congestion or information on the predicted travel time is generated based on the simulation results recreated using data synchronization technology, to calculate the optimum route guidance. There are various control policies avail-

Run simulations

- Collect mobility-related data (traffic and transportation) from urban areas in real-time, synchronize and update simulations

t=n
t=2
t=1

[
- Population measurement data (sensor data)
- Transportation system data (vehicular models, delay information, etc)
]
Real-time data

[
- Population measurement data (sensor data)
- Transportation system utilization data (number of people passing through ticket gates)
- Event data (number of attendees)
]
Past historical data

[
- Transportation system timetable
- Event security plans
- Environmental information (stadium, station, roads, etc.)
- Pedestrian models
]
External information

Real-time traffic simulation results
(spatio-temporal data)

- Enter external information or past historical data, and run learning and synchronization to accurately recreate the flow of people in simulators

Display information on predicted travel time covering the last-one mile

- Improves convenience for users through their travel patterns, and also encourages leveling out transportation system utilization with (1) Space-based (route) dispersion and (2) Time-based dispersion.

| Displays a comparison of transportation modes with services that can actually be used, based on the Travel time leaving Now in (1) Space-based dispersion. | Displays the changes in travel times for each departure time, based on the Travel time leaving Soon in (2) Time-based dispersion. |

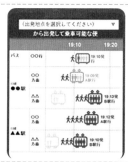

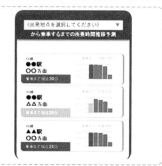

It will take time until I can board the train at XXX Station that I used to get here.
→ I'll go home via OOO Station instead of using XXX Station.

Attendees

Even if I leave for the station now, I'll get caught up in traffic which will take more time. If I go 1 hour later, after 8PM, it will be quicker.
→ I'll board the train after shopping at stores around here for an hour.

Attendees

Fig.3-1

able for settling on the optimum guidance, including guiding people, changing routes, changing road widths, changing signal patterns, and increasing the number of train services, and the number of potential combinations is immense. Yet running simulations for such a large number of potential combinations would take too long. Instead, we have developed control policy optimization technology that predicts from a small number of simulation results how good (travel time/degree of congestion/guidance cost) a route guidance that has not yet been tried yet would be, and then calculates the next guidance that should be simulated, to achieve efficient search results.

(3) *Behavioral change* requires improving convenience for users through their travel patterns, and encouraging leveling out the utilization of transportation systems with *Space-based (route) dispersion and Time-based dispersion*. We are aiming to alleviate congestion by leveling out the time and routes of transportation systems that attendees use, based on the results acquired with data assimilation and optimization. We achieve this by displaying multiple potential modes of transportation from the user's location, and predicting and indicating changes in travel time depending on each departure time, as well as providing incentives for users by suggesting what actions they should take to avoid congested conditions.

02 Virtual Reality / Augmented Reality

Closing in on humans' diverse sensations; Seeking the essence of reality

The market for Virtual Reality (VR) and Augmented Reality (AR) is expected to skyrocket by 86% annually from 2017 to 2022[1]. These technologies are already commonplace throughout the entertainment industry, and they are also being increasingly used in B2C business scenarios. An example is AR promotional campaigns run by apparel brands to try on clothes. Utilization of AR and VR is also being considered by B2B business throughout a wide range of industries. There are a broad range of future applications expected in the mobility, finance, real estate and medical fields, with examples including viewing three-dimensional renditions of industrial products or buildings during the design process, and administering remote treatment with the use of virtual surgery.

So what are the requirements for utilizing VR or AR in such an extensive range of applications? There are three topics that are drawing attention with current research into VR / AR: *Deepening presence*; *Speeding up processing*; and *Controlling motion sickness*. If we can overcome the challenges in these areas, VR / AR may evolve to become a technology that changes the way we interact with *reality*, instead of merely as a tool for the purposes of entertainment or virtual experiences.

Smarter simulation of complex senses, instead of just visual

The first topic, *Deepening presence*, means to enrich the *presence* of the subject to better resemble its true essence, by stimulating various senses instead of just the visual. When people hear the words VR / AR, the first thing they imagine is likely to be a head-mounted display. This is indicative of the way that many VR / AR systems developed in recent years control our *vision* to convey the presence of a subject. Yet there have been efforts to provide an experience that feels more realistic, through a combination of the sounds, acceleration, vibrations and other effects that are commonly used in games and simulators. As future advances in technology serve to stimulate a more sophisticated combination of senses including vision, hearing and touch, this presence may indeed become more enriched to feel even more realistic.

Speeding up processing and Preventing sickness

Next up, *speeding up processing* quite literally means accelerating the processing speed of the VR / AR experience to deliver something that feels even more realistic. Compared to actual scenarios, the image quality of VR / AR is severely hampered due to the bottlenecks from today's semiconductor performance and communication speeds. If processing speed can be increased with enhancements to algorithms or faster communication speeds with 5G and other advancements, the VR / AR experience will certainly benefit from higher quality images. An example of this is a 2016 paper published by Sebastian Friston et al. at UCL that focused on the topic of frameless rendering technology. [2] In the past, video needed to be loaded frame-by-frame. Frameless rendering was originally put forward by Gary Bishop et al. in 1994 as a means of overcoming bottlenecks caused by high latency. The research conducted by Bishop et al. used FPGA (Field-programmable gate array: reprogrammable hardware) to demonstrate its potential for faster rendering.

Yet the VR / AR experience will not become more enriched by simply increasing processing power. Another issue to address is the fact that there is the risk of getting sick from using VR / AR. The issue today is that people are unable to watch VR / AR for long periods of time due to the onset of sickness. Numerous studies are being conducted to address this shortcoming, including a 2018 study by Yue Wei et al. at the Hong Kong Institute of Technology in Ergonomics, which focused on situations where movement across the entire field of vision is used to give the user the sense of movement. The results of the team's studies indicated the possibility of reducing sickness by shifting the user's awareness towards their peripheral vision, instead of focusing their attention to their central vision. [3] There are a host of practical methods being developed regularly throughout the entertainment industry to reduce the sense of sickness, with the aim of enhancing VR gaming experiences.

Eliminating the barriers between the real world and the VR / AR world

NTT has been focusing its development efforts on VR / AR technology from an even greater range of perspectives. Some of the fruits of these efforts are already available in the entertainment and other fields, but we are not merely settling on entertainment as the goals of our endeavors. We have set our sights further afield, to create experiences where people are no longer limited by time and space to be at one with their natural environment, and also to be able to share those experiences as they like. An example is *Ultra-realistic Communication Kirari! 2.0*, which will use zero-latency technology for highly realistic images to overcome the delays that usually occur with remote transmission.

Meanwhile, we are modeling people, objects, the environment and other details of the real world, and merging them with Digital Twin Computing (DTC) information that has been recreated, predicted or expanded on using simulations that incorporate individual values and experiences. People might be able to experience the future worlds or societies generated with DTC by using VR / AR, through the senses of vision, touch and hearing just like in the real world.

And as more advances are made to *information collection and processing*, *real-time synchronous transmission*, and *realistic effects and reproduction* technologies, the very essence of reality will become better defined. This will significantly enhance the presence of subjects and in turn lead to a much higher quality that people can experience. These technologies are already being rolled out as sporting scenarios or events that blend traditional performing arts with modern technology, and there are bound to be more opportunities to experience these technologies in the future.

As NTT continues to push ahead with research into VR / AR, we also believe that *ambient* will become a vital keyword looking forward. This is because such technology will be increasingly present in the most ordinary of lifestyle environs, and people will be utilizing it without even being aware that they are. At the moment, there is a clear *barrier* between the real world and VR / AR world, as the technology requires dedicated devices that restrict the movement of people. We are aiming to create a world where VR / AR blends in naturally to everyday life without impeding on their activities.

These technologies not only serve to deliver new experiences to humans, but can also be utilized to supplement and boost human capabilities. As an example, we envisage being able to apply this technology to humans in much the same way as a *sixth sense*, and acquire information that cannot normally be detected by human sensory organs. In this way, humans may be able to make use of senses as if they were an extension of their ordinary bodily functions.

As research into VR / AR gathers momentum, technology will become available that offers a more sophisticated fusion of the usual senses like vision, hearing and touch. This is also crucial technology for creating the conditions where anyone—irrespective of level of IT literacy—is able to reap the benefits afforded to them by IOWN. But first we want to continue focusing our efforts on the various senses that people have, while also boosting research for bringing about even higher quality realistic results in the future.

[1] Allied Market Research, Global Augmented and Virtual Reality Market, 2014-2022

[2] Sebastian Friston et al.,"Construction and Evaluation of an Ultra Low Latency Frameless Renderer for VR,"*IEEE Transactions on Visualization and Computer Graphics*, Volume 22, Issue 4, pp. 1377-1386, Apr. 2016

[3] Yue Wei et al., "Allocating less attention to central vision during vection is correlated with less motion sickness," Ergonomics, Vol. 61, pp. 933-946, Jan. 2018). There are many efforts, including this paper, that are focusing on creating visual effects that do not create a sense of sickness.

Case Study 4

Kirari!, ultra-realistic communications
A viewing experience out of this realm

Kirari! ultra-realistic communications is capable of transmitting the *entirety* of competition spaces or live event venues over networks in real time to remote locations and makes it possible to experience a sense of immersion as if you were right there. *Kirari!* is a set of technologies that controls and processes media information and transmits it synchronously in *real time*. It faithfully transmits event spaces synchronously in real time to distant concert halls or performance venues, to create a groundbreaking ultra-realistic viewing experience.

Sophisticated elemental technology for vision and hearing
Kirari! comprises multiple elemental technologies as outlined on the following page **(Fig. 4-1)**.

The first element is *real-time image segmentation for any background*. This technology extracts only the subject areas in real-time from videos of sports arenas or live venues, without using traditional film studio shooting techniques like green screen backgrounds. The machine learning system derives high-precision contour and color based image extraction that can identify even the smallest differences in features between the foreground and background, which was not able to be distinguished using conventional techniques. Also it achieves more robust and high-precision image segmentation by semantic based segmentation using convolutional neural networks (CNN) and image feature detection using multispectral imaging.

Ultra-wide image synthesis technology and *Ultra-realistic immersive telepresence synchronous technology (Advanced MMT)* are technologies for synchronous transmission to remote locations without delays, by stitching together video feeds from multiple cameras covering a wide field of view around sports arenas or performance stages in real-time. This works by identifying seams in video frames that avoid moving objects, to prevent artifacts like image loss or stretching caused when moving objects cross a

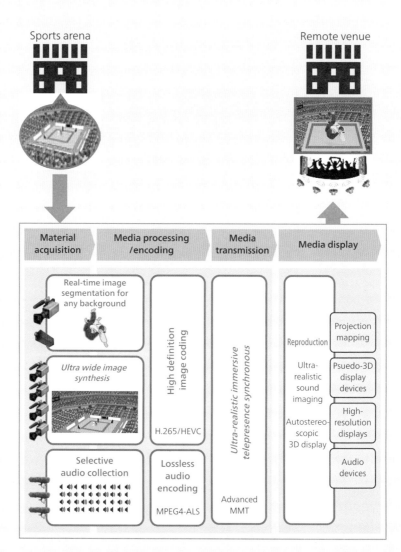

Fig.4-1: Kirari! technical layout

seam. The various media processing steps required for stitching and transmitting video are controlled as tasks using GPUs (Graphics Processing Units). This allows it to be used for live broadcasts with general-purpose servers, which eliminates the need for dedicated hardware system.

The spatial sound reproduction technology is also used to reproduce sound images at various positions around venues using a speaker array comprised of multiple speakers. The technology is based on sound field synthesis, and delivers special sound effects right up to the audience seats, thereby making it feel like the audience at a remote venue are cheering right nearby.

This is also enhanced with *autostereoscopic 3D display technology (Glasses-free 3D)*, which creates natural three-dimensional scenes. Autostereoscopic 3D technology enables viewing of 3D objects with binocular disparity on a screen without using 3D glasses by combining multiple projectors and a special screen. This special screen uses linear blending technology dubbed spatially imaged iris plane screen that changes the luminance ratio smoothly between adjacent viewpoint images as the viewpoint moves, enabling seamless motion parallax despite having significantly fewer projectors than conventional technologies.

Various ultra-realistic viewing experiences

So then, what visual experience does *Kirari!* actually deliver to viewers with the combination of these elemental technologies? **(Fig. 4-2)**

For instance, using *real-time image segmentation for any background* enables life size visuals of identified subjects to be projected with pseudo 3D display devices. And the shadows that those subjects cast are also generated automatically and displayed on the background, which gives the visuals a much greater sense of depth than ordinary 2D displays. When this technology is used to project sporting events, it creates such a level of reality that it feels like the athletes are competing right there in front of the viewer. Furthermore, by sensing the performers or athletes and using media processing, the current performer can be shown at the same time as one who was on stage moments earlier, to create a visual projection that would be impossible using conventional techniques.

And if pseudo 3D displays are extended in four directions *(Kirari! for Arena)*, the videos displayed on each surface is synchronized perfectly with the use of *Ultra-realistic immersive telepresence synchronous technology*. In this case, multiple people can

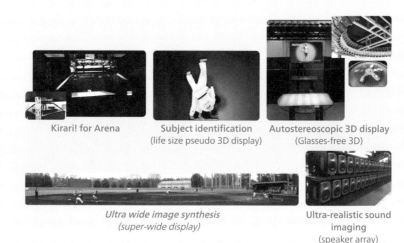

Kirari! for Arena

Subject identification
(life size pseudo 3D display)

Autostereoscopic 3D display
(Glasses-free 3D)

Ultra wide image synthesis
(super-wide display)

Ultra-realistic sound
imaging
(speaker array)

Fig.4-2: *Kirari!* Imprementation layout

walk around the display device and watch the action within from different directions.

With *Ultra-wide image synthesis technology,* super-wide images beyond the 4K/8K field of view and resolution are projected onto huge screens with multiple projectors, and audiences can view the events as if they were watching live at the actual venue. Videos displayed in full scale, life size dimensions creates realistic scenes with a level of liveliness and energy as if the athletes were competing right in front of the viewer.

Using *autostereoscopic 3D display technology* projected from multiple projectors in a 360-degree arrangement, large audiences are able to discover a new visual experience as if they themselves were immersed within high-reality 3D images of sporting events displayed floating on a large-size table.

The spatial sound reproduction technology based on sound field synthesis can control the distance of sound images, which is difficult with other sound reproduction technologies like surround sound reproduction. Audience members can experience the full scope of up-close sound effects, as if they were sitting with the audience at remote venue rooting for their team when cheers are reproduced by the technology.

We will be demonstrating *Kirari!* in action in the fields of sports and entertainment, as well as enterprise applications, as we aim to generate digital realms capable of creating and delivering natural-feeling experiences spanning far beyond spatio-temporal constraints.

03 Human Machine Interface
Recognizing the internal workings of human beings, to extend
senses and motion

As technology advances and a whole range of industries embrace everything *smart*, what changes lie in store for the *physical being* of humans? The Human Machine Interface (HMI) can be considered the key to broadening the scope of our bodily functions and extending them in a more natural manner. HMI refers to the technologies and systems that allow humans and machines (or mechanical devices) to exchange information with one another. The capabilities of HMI are predicted to increase into the future, in line with developments in cognitive science, neural science and other fields that escalate our understanding of the human mind and physical body.

In what way, then, is HMI likely to extend the bodily functions of us humans? Countless examples exist. In the construction industry or maintenance and servicing industries, humans will be able to control robots in remote locations as if they were moving a part of their own body. In the medical arena, patients will be able to undergo remote surgery accurately and safely via robots operated by skilled surgeons. There will eventually be a day where combinations of VR or AR are used for external control of micromachines inserted into a patient's body for performing surgery.

Voice-operated interfaces are already in use today to allow humans to control machines. Control based on eye tracking and other gestures have also become more commonplace in recent years. Greater development of brain-interfaces that make use of information sourced from brain activity may also become more sophisticated into the future. As research into HMI progresses, research efforts are continuously focused on two core topics: *Developing upstream input*; and *Reflecting cognitive processes*.

Developing upstream input and Reflecting cognitive processes
Until now, HMI has relied on detecting the way humans move—whether their

four limbs or their line of sight—or voice commands as forms of input. These types of movement actually occur when the brain sends electrical signals via the nervous system to each organ in the body. Another approach makes use of information sent to muscles or from the brain that controls the movement of muscles, instead of using input based on muscle movement (like fingers or voice) as in past systems. In this way, *development of upstream input* aims to detect activities at an earlier point *upstream* of body organs than previous inputs. Yet issues with noise and inaccuracy are introduced with such activity signals, which makes it difficult to develop an interface that operates smoothly.

Countries around the world are making moves in this field of research, including the 2018 joint research project headed by Enhao Zheng of the Chinese Academy of Sciences and Qining Wang of Peking University, et al.[1] In this work they developed an interface capable of recognizing forearm myoelectric signals without contact with the human skin. There is also research into devices inserted directly into the brain or skull. This is highlighted by research efforts being made in the United States for an implant-type neural interface that connects computers with the human brain.

Meanwhile, *reflecting cognitive processes* is aimed at developing interfaces with minimal errors. Avoiding errors is difficult with HMI, yet as HMI becomes more prevalent in line with an increased understanding of human cognitive and thought processes into the future, it may be possible to develop interfaces that mitigate errors.

Natural, intuitive extensions of the human body

With existing HMI, it was difficult to avoid the feeling that humans were actually *using* devices. Yet looking forward, there will likely be a time where humans use devices or robotics in a natural manner. At NTT, we are focusing our efforts on expanding on capabilities of humans in a natural, intuitive manner through the use of HMI.

One example that we are pushing forward today is the development of *ambient awareness technology*. This comprises devices equipped with multi-modal interfaces for sensing in real-time a whole host of information—from the surrounding environment, biological information and even sensory information that humans can acquire from their five senses—while at the same time providing information for natural interactions without impeding on human activities.

And more than just devices, a concept dubbed *ambient assist technology* supplements and boosts human capabilities while working in concert with the surroundings, as part of efforts to achieve an environment that further amplifies the five senses of humans through HMI. We will be turning our attention to better understanding the way people function internally, including all of our emotions, and develop technology that we can use to update this HMI.

HMI is indeed the one area where technology and *humans* have the most intertwined relationship. Deepening our understanding of humans and creating new HMI will serve to further expand the capabilities of humans. HMI will no doubt deliver an environment in which everyone everywhere—irrespective of level of IT literacy—will be able to reap the benefits of technology.

[1] Paper published in July 27, 2018 *Frontiers in Neurorobotics* (Enhao Zheng et al., "Forearm Motion Recognition With Noncontact Capacitive Sensing").

Case Study 5

Point of Atmosphere
Toward a device-less future

At some point in the future, we envisage that there will eventually be a time when there is no longer any need to pay direct attention to visible devices like smartphones. Indeed, there will be more and more devices around us that keep a close watch over our lives. For example, various ICT devices in rooms operate in concert to convey to users in a natural manner that it will be raining today, without them having to look up weather forecasts: raincoats hanging on walls may wiggle, or a visible effect may make the floor appear wet. We call such an environment that prompts people, *Point of Atmosphere (PoA)*, which enhances DX (digital transformation) as people move and interact seamlessly with their surroundings, without interfering with their primary activities.

Ideal services that *naturally* assist people

Imagine standing in front of a mirror, which shows an overlay with the best outfit to wear on that day. Until now, services like these have been portrayed numerous times as a glimpse of life in the near future. *PoA* takes this approach a step further as it aims to create a vision of a person's future. Examples of this in practice include viewing a healthy version of someone in the future as a result of proper health and hygiene management. In contrast, a more haggard-looking version of the person after years of unhealthy living might serve as a warning that could be useful for improving awareness of health management. The aim is to create an atmosphere where useful information is conveyed naturally from surrounding items, even when doing errands outside.

While *PoA* is likely to blend in seamlessly with our daily lives in the real world, it is also expected to play a role in a broad range of industrial sectors. Examples in the field of transportation include being provided with train departure information without having to directly look up timetables, or alerting people of dangerous situations by triggering greater awareness when cars are approaching. Another potential application is predicting in advance when severe disasters might strike, and activating transportation system fail safes (remote shut-down) or switching over to public cooperative driving zones.

And in the medical field, a patient's symptoms or emotional state may be observed to encourage behavior acceptance and seamlessly improve their lifestyle as a way of preventing illnesses. If surgery is required, a diagnosis may be made remotely and robotics used to administer treatment as if a well-trained doctor was performing the surgery. In such cases, the patient's medical records can be examined to predict the progress of surgery being performed and identify any risks. This could help to ensure that surgery is performed without any mistakes being made.

In the entertainment industry, audiences may be able to watch sporting events with the same level of excitement as if they were right there at the sports arena. And the sense of adrenaline that athletes feel might be able to be experienced by the audience tuning in. With performing arts, today's actors may be able to share the stage alongside famous actors from the past as part of special performances that transcend spatio-temporal constraints. The same concept could apply at restaurants. Diners could instinctively learn more about the local regions of ingredients used in dishes, or the ingenious ideas employed by manufacturers or chefs—all of which would make for a more enjoyable dining experience.

The technologies underpinning *PoA*

There are a range of technologies that lie at the heart of *PoA*. (1) *Ambient awareness technology* that seamlessly blends people with their surroundings to create natural behavior beyond spatio-temporal constraints (regardless of time or place); (2) *Ambient assist technology* that supplements and assists human activities by coordinating the environment and devices; and (3) *Cybernetic UX technology* that seamlessly coordinates and unites humans with devices and robotics based on the mechanisms of human activities.

(1) *Ambient awareness technology* forms the basis for natural behavior when humans interact with information. Various sensors installed throughout the environment, and devices that naturally convey information to the five senses of humans, are melded seamlessly into the ambient surroundings to operate and coordinate together

in concert. (2) *Ambient assist technology* extends the ability of humans, and can also supplement any bodily functions a person may have lost. Expanding human capabilities and functionalities can lead to new discoveries never experienced before. (3) *Cybernetic UX technology* draws on basic human instincts, and guides people so that they can live a better life.

There are two processes running in the background to make these technologies possible: AI technology for thorough and extensive analysis and assessment of various forms of information; and the Digital Twin Computing (DTC) platform for transmitting this information to cyberspace where it can be stored and managed. The DTC platform creates a virtual parallel world in cyberspace in which everything within runs simultaneously alongside the real world. It is possible to expand the capabilities of people within this parallel world. An example to illustrate this concept may be someone who has difficulty conveying their thoughts due to language barriers in the real world. In the parallel world, they are able to converse naturally, which makes it a valuable tool for broadening the scope of their activities.

To ensure that everyone regardless of their IT literacy can take advantage of the benefits of this parallel world, we will be advancing our research and development efforts while utilizing sensing and media processing technologies.

Case Study 6

Deformation Lamps and *Hidden Stereo*
Presenting information using optical illusions

NTT has a long history of studying the characteristics of human perception. We are advancing our research efforts to better understand how and why humans are able to see objects, listen to things, or experience the sensation of touch. These efforts aim to better explain fundamental perceptual characteristics, and more recently there are an increasing number of examples where scientific findings from research into perception are being applied as technology for conveying information. The following are several key examples that help to illustrate this.

Deformation Lamps light projection technique based on how movement is perceived

The technology used for projecting images or videos onto the surfaces of objects with a projector to alter the way the actual target surfaces appear is known as projection mapping, and is widely used today. NTT developed a new way of using this projection mapping called *Deformation Lamps*. This is a light projection technique where only information related to movement is added, while retaining the tones and patterns of the projection target surfaces. Deformation Lamps creates an optical illusion by projecting light and dark moving patterns onto a static target. This makes the target itself look like it is actually moving. Take a look at the portrait of Johann Sebastian Bach. The visual effect makes it appear as if he is smiling all of a sudden **(Fig. 6-1)**.

So how do Deformation Lamps make it look like the target is moving? Video c below projected by Deformation Lamps is a video generated from the differences between the original static image a, and image b which has movement artificially added to image a by a computer. Here, the differences are computed in terms of luminance intensity between the two images. Projecting a video comprised of these differences onto a static target makes the target look like it is moving in the direction of those differences (video c). The most interesting aspect here is that the projected video con-

Fig.6-1

a. Before projection   b. Projected video   c. After projection

tains no color data at all—the Deformation Lamps technique is even able to add motion effects to targets containing colors. The reason for this is that the human visual system is quite sensitive to light and dark movements, but less sensitive to movements of color patterns without luminance variations. It is thought that movements in brightness are also perceived by human observers as movements in color.

It goes without saying that projecting light so that the way the original image is viewed is completely replaced with the version that has moved will achieve the same effect without using optical illusions. Yet such a method would require bright lights to be projected in dark environments, which would significantly diminish the natural look of the target. Armed with our scientific understanding of human vision, Deformation Lamps allows natural looking movement to be created with the minimum amount of light needed for detecting movement.

Deformation Lamps are already available for practical applications, and they are increasingly being used as a new way to express movement in the fields of advertising and the arts.

Hidden Stereo video generation technology
viewing 3D and 2D simultaneously

Humans use the spatial disparity (parallax) of information received by the left and right eyes for depth perception. The most popular types of 3D videos comprise a pair of two offset images that are displayed together on a screen. Viewing this display while wearing 3D glasses separates the images so that each is viewed by the left and right eye respectively. Yet when viewing conventional stereo images without the 3D glasses, a viewer sees two images that appear blurred, which results in an uncomfortable visual experience. This meant that people wanting to view 3D videos, and people who want to watch videos casually without wearing glasses, were unable to watch videos in the same place.

To overcome such issues, NTT developed the stereo image generation technology *Hidden Stereo* which utilizes a similar visual effect to Deformation Lamps. Just like Deformation Lamps added patterns to convey the sense of motion, Hidden Stereo adds patterns to generate images with parallax. More specifically, for a single original image, a disparity-inducing pattern is added to generate an image for the left eye, and the same pattern is subtracted to generate an image for the right eye **(Fig. 6-2a)**. The disparity-inducing pattern is designed carefully so that adding/subtracting the pattern produces offsets in opposite directions from the original image. This results in parallax between the left and right images, to create depth perception when viewing them through 3D glasses. Viewing these images without wearing glasses results in the images overlapping, the disparity-inducer components canceling each other out, and the original image is visible as a clear 2D image **(Fig. 6-2b)**.

The human visual system is able to accurately detect the displacement (phase shift) in apparent brightness between the left and right images, and depth perception is created using this characteristic. The disparity-inducing patterns are actually added to and subtracted from the original image to generate phase shift. While these stereo images differ to physically correct images of scenes viewed from two different viewpoints, they are almost identical to the human visual system. Hidden Stereo can be viewed as a technology developed based on our understanding of the human visual system, with videos stripped of visual information except for the bare minimum required for humans to gain a sense of depth perception.

Utilizing Hidden Stereo allows users to enjoy a 3D and 2D hybrid viewing experience at movie theaters or at home with 3D televisions. Another option is to project only disparity-inducing patterns like Deformation Lamps to incorporate 3D vision onto surfaces in real space.

NTT is setting its sights on applications developed based on data projection technologies like these, and will also play a role in advancing fundamental scientific knowledge into the future.

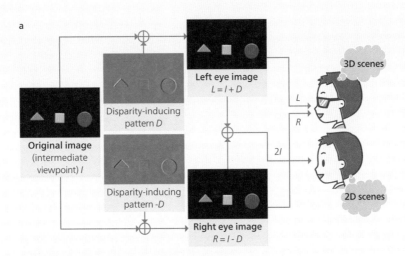

a

Original image
(intermediate
viewpoint) I

Disparity-inducing
pattern D

Disparity-inducing
pattern $-D$

Left eye image
$L = I + D$

Right eye image
$R = I - D$

L

R

$2I$

3D scenes

2D scenes

b

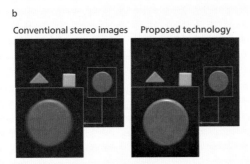

Conventional stereo images   Proposed technology

Fig.6-2

04 Cybersecurity

Protection across a broad scope with preventive and adaptive solutions

There is a growing risk of cyberattacks becoming more sophisticated in recent years, and their effects are becoming more widespread. As such, the importance of having capable security measures in place is increasing at a faster rate. Today more services are available in the cloud, and there has been a marked increase in the number of devices connected to networks thanks to IoT. All of these mean there are more targets open to attack, and corporations can no longer afford to merely keep one step ahead of cyberattacks—preparation is required to guard against attacks from a whole range of different approaches. Because of the increase of such needs, the global cybersecurity market is predicted to grow at a high rate of 11% per annum through to 2024 [1].

In the past, cyberattacks primarily referred to data manipulation or breaches, but as networks and IoT equipment become increasingly sophisticated, attacks against them have a higher risk of triggering major accidents in the real world. If connected cars, drone delivery networks or medical treatment systems at hospitals were to be hacked and taken over, the resulting accidents could be life threatening and have devastating consequences. Rather than just being able to withstand such attacks, it is becoming increasingly important to have preventive security measures in place to minimize the risk of such damage.

Developing preventive, adaptive safeguards

To address these types of concerns, research efforts into preventive protection are on the rise in the cybersecurity sector. Let's explore this further using the well-known DDoS (Distributed Denial of Service) attack. This is one type of attack that is launched over a network, in which target computers are flooded with excessive processing load from numerous machines in an attempt to interrupt services. The methods used to protect against such attacks in the past included decentralization of servers/networks and packet filtering. Anti-virus and similar software packages were used to prevent vulnerable terminals from being taken over and utilized for DDoS attacks, while network scanning and other tools were deployed to detect devices that lacked the necessary security. Yet the cyber criminals employed various methods to get around such security measures: increase the number of attacking nodes; exploit unknown vulnerabilities; or phishing techniques to gain access to devices with attacks via users.

Preventive safeguards are required in order to overcome these workarounds. With preventive safeguards, system logs and other data are analyzed using machine learning or similar techniques, which enable the method of attack and degree of impact to be identified for a quicker response. This helps to lessen the amount of damage in preparation for attacks, and allows a more adaptive response to be taken.

In 2015, Stephanie Gil et al. at Massachusetts Institute of Technology demonstrated the detection of *Sybil attacks*, in which a single device can control reputation systems by creating multiple pseudo device identities. This was shown to be possible by identifying physical position data of a device by analyzing wireless network signals [2]. They suggested that it was possible to detect the spoofing device before an attack actually takes place.

And in 2018, Abebe Abeshu Diro et al. at La Trobe University proposed attack detection systems using deep learning that could be distributed over large scale networks [3]. This study aimed to increase protection performance by utilizing deep learning, while retaining the scalability of attack detection systems. This type of research broadened the potential of employing high-performance, automated preventive protection to monitor large scale networks such as IoT networks used throughout smart cities.

Broader protection, from the IoT, mobility, social infrastructure, and even people

Security is of extreme importance in the IOWN concept that we envisage at NTT. While high-speed mobile networks based on 5G will help create a more intelligent world, it is also likely to bring with it a new range of cyber threats as various devices gain an online presence where they had no network connection in the past. With the possibility of such risks, we are focusing our efforts on creating advanced security systems while applying the technologies we have developed thus far. There are three areas that we are focusing on in particular: *Expanding protection for more devices*; *Developing preventive solutions*; and *Cryptography.*

The first, *Expanding protection for more devices*, aims to accurately detect and respond efficiently to cyberattacks as a way of protecting against threats to people as well as IoT devices and mobility that are closely connected to the real world, in addition to existing IT sectors. The chief focus of protection against attacks closely connected to the real world includes security technology for mobility networks in the leadup to the imminent era of connected cars, as well as security solutions in the field of industrial control systems such as plant equipment.

An illustration of this is with our collaboration with Mitsubishi Heavy Industries, Ltd., where we are advancing development of attack detection and response technology for industrial control networks used for social and other infrastructure. It is hoped that this technology will form part of security systems in more public sectors such as power stations and public transportation systems. Changing security rules in real-time for each protected device helps to identify faults quicker. This will make it possible to maintain availability while also being able to respond swiftly to unknown cyberattacks.

We are also advancing research into cyber physical security technology, which is the security technology used to support these new efforts. This technology also applies directly to the aforementioned actual society (physical space) that we live in, and protects against attacks that have the risk of affecting life or physical property. With this in mind, we are aiming to achieve detection and response with an even higher level of immediacy and accuracy compared to technologies utilized in solutions for conventional cyberattacks. As cyberspace and physical space also become more closely linked, the overall systems running in them increase in complexity. As such, the methods employed for attacks are also predicted to increase in complexity and sophistication. We are working to clarify the mechanisms that may be used for attacks and damage to such complex systems.

There is ongoing debate about the stale of security technology that takes into consideration use by humans, dubbed usable security and privacy, as part of studies into protection against attacks targeting people. Instead of providing protection for only ICT systems, we also focus on the people that use such ICT systems, and understand their level of awareness and behavior. We are attempting to enhance security in a multifaceted manner by ensuring that security can be understood and used by anyone.

Detecting signs to develop solutions in advance

For the second task, *Developing preventive solutions*, the activities we are implementing include: (1) Analyzing and addressing methods of attack; (2) Detecting and addressing signs of cyberattacks; and (3) Identifying and addressing malicious sites.

The first of these makes it possible to detect attacks that do not trigger conventional anti-virus software, as quick as possible. This is achieved by collecting data on the most recent malware samples or attacks that exploit vulnerabilities using *honeypot*, or baiting, systems to swiftly create an IOC [4].

The second activity aims to develop more advanced countermeasures by collecting, analyzing, and deploying to a broad range of sources, including information on the dark web via NTT-CERT [5] or other non-public information cooperating sources [6].

The third activity helps to identify malicious sites [7] that disappear quickly by probing sites [8] to create a blacklist almost in real-time. These efforts ensure a greater level of detection precision than ordinary blacklists.

As a mega-carrier responsible for handling such vast quantities of data, it goes without saying that NTT views cybersecurity as a field of the utmost importance. This not only serves to meet the needs for keeping people safe and secure, but is also essential for developing even more stringent cybersecurity systems required for maintaining protection during the soon-approaching smart world era.

Cryptography underpinning preventive, adaptive safeguards

The third area of focus, Cryptography, underpins the aforementioned *Expanding protection for more devices* and *Developing preventive solutions* as a core technology for protecting data and making sure it is used safely. We are advancing research into cryptography with a view to the computing environment 30 years from today. This can be demonstrated with our research into advanced cryptographic theory. Examples include post-quantum cryptography, cryptographic methods designed to ensure security during the era of quantum computers that will no doubt be developed in the future, and cryptographic program obfuscation to protect software intellectual property by making it impossible to analyze processing details of an application.

We are also developing technology that protects and utilizes data securely at end-to-end of data generation, distribution and analysis. We are researching data protection technology and privacy protection technology that ensures a platform for safe and secure data distribution, such as encrypted communication protocols that allow uniform secure communications used by low-functionality devices to high-performance servers; secure multi-party computation that enables data analysis while keeping data encrypted; and anonymization technology that processes personal data to prevent identification of individuals.

[1] Grand View Research "Cybersecurity Market [2014-2024]"

[2] Paper published in *Autonomous-Robots* Vol. 41 (Stephanie Gil et al., "Guaranteeing Spoof-Resilient Multi-Robot Networks).

[3] Paper published in *Future Generation Computer Systems* Vol. 82 (Abebe Abeshu Diro et al., "Distributed attack detection scheme using deeplearning approach for Internet of Things").

[4] Definition file used for detecting behavior suspected of infectious activity.

[5] A specialist team that responds to computer security incidents within the NTT Group.

[6] Non-public information related to cybercrime.

[7] Includes C&C servers that issue attack commands to infected devices.

[8] Inspecting sites in detail.

Case Study 7

Piper, botnet detection framework by analyzing large-scale data flow
Facing internet-based threats head-on

While the rapid uptake of the internet has led to greater convenience for users, the frequency of cyberattacks utilizing the internet has also increased in recent years. Cyber criminals create a platform on the internet called a botnet, which can be used to perform DDoS attacks, network scans or a variety of other malicious activities. This is where *Piper* plays a role—we developed this botnet detection framework to detect a range of attacks like these and maintain a secure internet environment.

The growing botnet threat with the rapid spread of IoT devices
Botnets refer to the networks connected to by devices (bots) infected with a range of malicious software, which allows bots to receive instructions to attack from a C&C server over the internet. Recent cases have seen infections spread chiefly through vulnerable IoT devices, leading to botnets that are becoming larger. In turn, this can result in DDoS attacks on a much larger scale that present a greater threat to online service providers.

Botnets are constantly changing as they themselves grow and develop to avoid being detected, and the methods used to do so become more complex with time. As botnets increase in size and complexity, it is vital to gain a full understanding (Fig. 7-1) of the activities of these botnets in order to stop them launching attacks.

To face the threats that these botnets bring head-on, NTT is coordinating with its group companies overseas for local research and development work on *Piper*. Botnets tend to hide their communications within large quantities of ordinary communications, which makes it difficult to detect them by simply using detection rules. To overcome this, we are studying how to apply machine learning to detection.

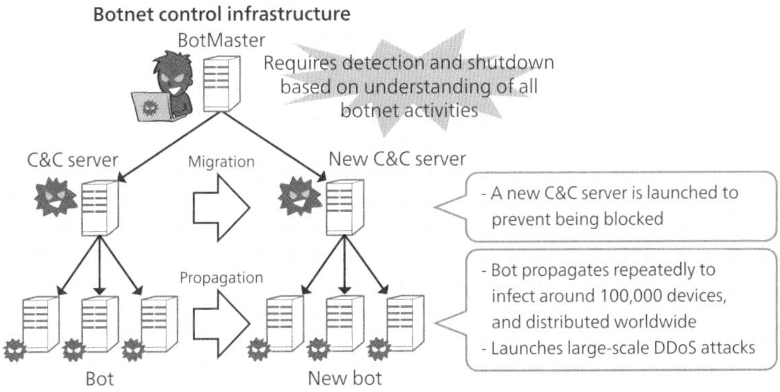

Botnet control infrastructure

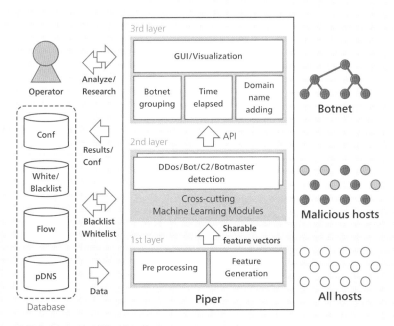

Fig.7-1: Overview of botnet activity

Fig.7-2: Schematic of Piper functions

Three layers of protection

Piper is configured with three layers **(Fig. 7-2)** that allows machine learning analysis of large-scale networks to be performed in near real-time.

The first layer involves identifying feature values from traffic flow data to achieve high-speed traffic processing, while also extracting feature values backed with NTT's expertise in botnets. The second layer features multiple machine learning detection modules, each of which uses algorithms with excellent capabilities for interpreting detection results, allowing the validity of those results to be verified. The system is highly-scalable, and individual modules can easily be added or changed, or updated with new machine learning modules as required. The third layer is made up of modules for visualizing the detection results. The core components of detected botnets are displayed as a graph to provide a comprehensive, layered overview of the botnet, including the routes of each core component like bots and C&C servers.

Here is a more detailed look at each layer. Information that can be acquired from traffic flow data is mainly limited to the 5-tuple (source IP, destination IP, protocol, source port, destination port), however *Piper*'s first layer extracts 500 dimensions or more feature values from the flow data. Of these, a large number of feature values have successfully been extracted from multiple perspectives.

For features associated with communications size, for instance, *Piper* is designed to detect bots that send packets of a fixed size to regularly test communications with its C&C server. And for features based on geographical distance, *Piper* is designed with the fact that communications between botnets have a poor correlation with geographical location. All feature values are updated with NTT's knowhow on botnet behavior, with test results indicating that botnet core components can be detected with a high level of precision compared to existing studies.

During the course of ordinary security operations, detected results provide an understanding of *why such activity was detected*, so that the appropriate solution can be devised after checking for the possibility of false positives or degree of impact. Conversely, detection results based on machine learning do not necessarily mean that such information is added. Even if detection results are correct, it is not always easy to determine which feature values were effective for detection, depending on the type of machine learning algorithm.

This is where the machine learning method used by *Piper* provides detection information that can be applied during operations. This is possible with the capability of algorithms used to interpret the results, including *Semi-Supervised Learning* and *Unsupervised Learning*.

In the third layer, the results detected with machine learning are used to visualize

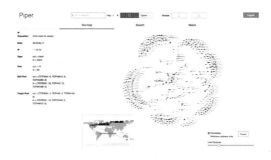

Fig.7-3: Example of Piper visualization

(Fig. 7-3) communication conditions as a graph to gain an understanding of what the level of botnet activity is, or what degree of impact it may have on actual services. From the visualization, nodes that are suspected of being activities of new botnets allow for additional analysis by drilling down the results.

As demonstrated above, *Piper* is capable of observing the activity of botnets. To streamline use of *Piper* in security operations, we will be pushing forward with further studies into methods for estimating the impact of detected botnets, or methods for actively observing whether or not botnets are actually in action.

LRR, platform for collaborative cyber defense
Preparing defense for constantly evolving cyberattacks

Damage caused by cyberattacks is becoming a major social problem recently, and most of this damage is caused by malicious programs called malware. More malware is constantly being discovered, and the methods used by cyber criminals are becoming more sophisticated on a daily basis. To ensure that protection against such attacks can keep pace with the attackers, sites where such attacks occur demand the latest cutting-edge technology to be available as quickly as possible. The number of cases dubbed targeted attacks is also increasing yearly, and solutions call for advanced technology to be applied. Targeted attacks are attacks aimed at a specific organization, which makes it difficult to track the state of attacks. To develop technology to protect against such attacks, collecting some form of data or knowhow only available at that organization becomes essential.

Responding quickly to cyberattacks

With these circumstances in mind, NTT is focusing on developing *LRR*, a platform for collaborative cyber defense that will form the foundation for *developing the latest cutting-edge technology as quickly as possible*, and *utilizing data or knowhow acquired from sites*.

LRR is equipped with a range of powerful tools: a URL inspection function useful for detecting and analyzing malicious websites that infect devices with malware when accessing them; domain name analysis function for classifying by degree of maliciousness of domain names or categories; and IOC (Indicator of Compromise) function that provides information that can be used to identify devices infected with malware.

These functions have been developed through a combination of the fruits of NTT's research and development efforts. Supplied with a GUI (Graphical User Interface) and API (Application Programming Interface) so they can be used straight away by operators or systems, one of the highlights of these functions is that they can be rolled out for practical applications with minimal delay.

Data from actual sites protected with *LRR*, as well as inspection requests and feedback from security analysts, and other stored information is also used during the R&D process to help improve *LRR* detection accuracy and enhance usability.

URL inspection and domain name analysis

Let's take a moment to explore each of the functions available with *LRR*. One of the most common methods of cyberattacks come from websites that have been modified for malicious intent. To prevent such attacks, the SOC (Security Operation Center) needs to monitor websites to check if they have been modified, and if so, to identify where they have been modified as soon as possible. Yet the methods used for attacks are increasing in complexity in recent years, which makes it difficult to identify such methods. To further complicate matters, identifying malicious sites requires manually analyzing huge volumes of HTML files and scripts, all of which takes an immense amount of time and resources.

To combat this, the URL inspection function applies multiple R&D technologies like *honeypots*, *machine learning* and *content analysis* to automatically identify websites that have been modified for malicious purposes. A GUI analysis tool is provided for SOC analysts to help find which sections of a particular website have been modified. The GUI analysis tool provides a visual map of how content is related and the way sites are structured, and also highlights malicious content that has been identified automatically, which assists operators for analyzing *why* a *specific section of a site* seems suspicious.

Domain names are an essential aspect of how we use the internet today. More than simply being used to browse the internet as you would normally, domain names are also employed for malicious purposes as necessary infrastructure for performing cyberattacks. Cyber criminals prepare new domain names on a daily basis for distributing their malware. More specifically, however, they also use domain names for a range of other illicit purposes, like those that appear similar to official services as part of phishing attempts to trick users, or for operating C&C (Command and Control) servers that send instructions to malware to launch DDoS attacks, send spam emails or otherwise steal unauthorized information.

The domain name analysis function assesses the degree of maliciousness of domain names from multiple perspectives using domain name repetition technology, to identify malicious domain names being used by attackers. Domain name classification technology is also used to verify the background and circumstances under which a domain name was generated, to present objective solutions that should be implemented to combat each malicious domain name. These technologies are also developed using NTT's proprietary store of security intelligence.

Creating infection trace patterns and analyzing malicious websites

While it is becoming more difficult today to completely shut out infections by malware due to targeted attacks and other means, indicators of compromise (IOC)—traces of infections remaining on hosts—can still be found with a vital technology called EDR (Endpoint Detection and Response).

NTT is focusing on research and development of automatic IOC generation technology with superior detection capabilities that can be applicable to EDR products. The process in detail first involves using NTT's state-of-the-art malware analysis technology to analyze collected malware samples. The malicious behavior contained within the malware is then fully extracted to create the patterns of the traces of specific behaviors of the samples. This is then used to generate NTT's proprietary IOC with superb detection capabilities.

The methods used by cyber criminals are becoming more sophisticated on a daily basis, and NTT is channeling its efforts into research and development of numerous other forms of protection to tackle these in addition to the technology outlined above. An example of this is collecting and applying intelligence on malicious websites that launch attacks by targeting the psychological vulnerabilities of users instead of systems. *LRR* will eventually be equipped with these new technologies in order to hasten the practical applications of the fruits of these cutting-edge R&D efforts.

05 Information Processing Infrastructure

Infrastructure for processing real world events in real-time

As technological innovation makes leaps and bounds throughout so many fields like AI and IoT, so becomes available many new ventures that were not possible in the past. While the advancements in software technology are impressive, hardware technology capable of supporting such advancements is essential for developing practical applications. Today, development of information processing infrastructure capable of running a broad range of advanced technologies is ramping up all around the world. The market for AI chips is one that is experiencing significant growth, with the market size predicted to expand at a rapid pace of 40% annually through to the year 2025 [1]. In a similar vein to AI, applications harnessing IoT are becoming increasingly common in a range of sectors, and the significance of information processing infrastructure is also growing.

With the rapid growth of the AI chip market, a range of AI devices that are aware of their surroundings while operating in the real world and capable of taking action autonomously are likely to be installed. These AI devices are initially likely to be single units that are able to take action based on the conditions around them. Yet there is a limit to the scope of conditions that single devices can monitor and act upon. Eventually, multiple AI devices will no doubt work in concert communicating with each other in real-time via high-speed networks.

By extension, if advanced AI devices interacting with the real world are able to connect and operate via networks, all manner of real world events and incidents could be interpreted and distributed as data. This means action can be taken at the most opportune time across society as a whole, without being limited to any single device. Picture, for example, an object that has fallen onto a road. Not only will self-driving cars be able to serve to avoid the object, but they could also notify others around them of the obstruction ahead. Vehicles receiving the notification could alter the route to their destination.

To make a world like this a reality, the infrastructure for converting various real world events (condition of things, events that are occurring) to a digital signal and distributing them will be needed. At NTT, we are advancing our R&D efforts for making such infrastructure a reality.

Improvements to computing efficiency with photonics-electronics convergence technology in the Post-Moore's Law era

Digitalization of events occurring throughout the real world could be achieved with advances made to sensor technology including cameras, as well as technology for analyzing sensor output data with AI. As the specific types of events to be covered depend on each individual case, AI analysis will be required for the corresponding sensor output. This means that computing infrastructure will be required to handle the AI analysis tasks for such vast quantities of computing data. Yet performance improvement by semiconductor miniaturization is fast approaching its limit', and it is becoming difficult to increase the computing efficiency of servers much further beyond the current level **(Fig. 1)**.

Photonics-electronics convergence technology is considered the key to

Fig.1: Computing challenges [2]

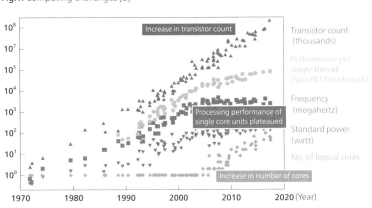

Source: https://github.com/karlrupp/microprocessor-trend-data

overcoming this barrier. Until now, data transfer paths between devices within a computer was achieved by electronic means, and physical limitations meant those transfer paths could not be increased in length. Difficulties in transfer control also made it hard to connect any more devices. Photonics-electronics convergence technology allows the data transfer paths between devices to be replaced by light, which maintains current data transfer rates while also allowing transfer paths to be extended further. This means more computing devices can be connected at higher transfer efficiencies to create computers with an immense boost in computing performance **(Fig. 2)**.

And if light can be used for data input and output of devices with photonics-electronics convergence technology, it will be possible for data to be transferred by light directly between devices instead of via their NIC (Network Interface Card)—a major advantage for distributed computing consisting of multiple computers. The result would be shared computing resources across different racks or even data centers, which would bring a major boost to computing efficiency throughout the entire infrastructure.

Fig.2: Data transfer rates and distance

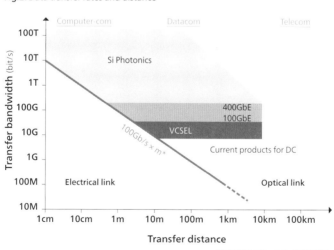

Krishnamoorthy, et al., JSTQE2011

Just like the development of transistors made pocket-sized radios with much lower power consumption possible, photonics-electronics convergence technology is likely to make computers and the cloud much more compact.

To bring about this type of innovation to computing infrastructure, NTT is moving ahead with research into *optical interconnect technology*, *distributed optical high performance computing technology* and *high-density AI inference techniques*.

Achieving second/sub-second order data flow

In order to respond swiftly and efficiently to changes in conditions occurring in the real world, data needs to be able to flow from where an event is unfolding and delivered to where it will be analyzed, on a second/sub-second basis. Yet with immense data volumes coming from so many locations, achieving this flow of data is no easy task.

A perfect illustration of this is with traffic. In the future, information (such as position, speed and destination) from millions of connected cars driving on the road needs to be updated, examined and collated in real-time. If the precision of positioning tracking information used by GNSS (global navigation satellite systems) increases, the update frequency will also need to be increased by several orders of magnitude to match. Processing the updates made to such immense volumes calls for systems capable of handling the geographical trends of the traffic flow itself, while also distributing the processing load in an efficient manner. NTT is developing *real-time spatio-temporal database technology* to address this issue.

As another example, real-time population heatmaps of commercial buildings, or temperature and humidity maps are available with useful analysis apps for not only the owner of those buildings, but also distribution centers, air-conditioning control companies, security companies and various other service providers. In this example, data is received from each commercial building and as such is spread over a wide area. Analysis apps are also used by numerous service providers operating in different locations. A new data transmission framework will be essential to ensure that data sent from various different locations is sent to various different analysis locations without any delays. NTT is developing the *iChie data hub* to answer this challenge.

Resolving combinatorial explosions with Ising model computing

So many events occur around the real world all the time. If data on those events can be generated and used for analysis at any location, detection, prediction and optimization that were difficult to achieve may indeed become possible. This will require prediction and optimization computing based on many input variables. As the number of available variables increases, generating models for detection or predictions, and optimization calculations become exponentially difficult.

A massive collection of data needs to be collated to determine the optimum routes and generate the route to take. For example, alleviating traffic congestion by optimizing the route of all moving cars or operating on-demand shared bus services requires the positioning information and destination of vehicles in the area, number of vehicles in each lane on the road, information on nearby construction work, information on nearby events and other traffic-related data, level of demand for transportation by users (in the current location and destination), desired arrival time, positioning information of shared buses, and so much more. The number of potential combinations of shared riding groups or routes is overwhelming, making it difficult for computers of today's technology to determine the optimum combination. To address this bottleneck, NTT has taken a different approach to existing computers and is developing the *LASOLV* computer (CS13) that uses a special laser oscillator called an Optical Parametric Oscillator (OPO).

There are still many issues that need to be addressed for information processing infrastructure. The limitations that current technology is facing need to be overcome in order to achieve the world we are envisaging with AI and IoT. Thus this new information processing infrastructure that we are trying to create will become the mainstay *infrastructure* that will be essential for supporting a smart world.

[1] Allied Market Research, Global Artificial Intelligence Chip Market, 2018-2025
[2] Data and diagrams up to 2010 sourced from M. Horowitz, F. Labonte, O. Shacham, K.Olukotun, L.Hammond, and C. Batten. Data and diagrams from 2010 to 2017 sourced from K. Rupp. These diagrams were supplied based on Creative Commons License 4.0. Translations and comments have been added to the original diagrams.

Case Study 9

AI optical interconnect technology
Playing a role in the post-Moore's Law era

There will come a time when there are billions, even trillions, of IoT devices connected to networks. The volume of real world data being acquired will simply be incomparable to today's level, and applications will be developed that will deliver wonder and excitement that we could simply not fathom in the past. To achieve this vision, computing power that far exceeds what was thought possible will be required for processing this immense amount of data in real world time. The optical communications technology that NTT has developed can be applied to create optical interconnect technology, which will be the key for delivering the advanced computing capability that will be required.

Distributed processing massive volumes of data at high speed

With limitations in power density as we approach the extreme end of Moore's Law, the era where the performance of single processors continues to increase every year in line with LSI process miniaturization is soon coming to an end. *Post-Moore's Law technology* will be essential for moving beyond this barrier and harnessing computing power that exceeds anything we have today, without relying on the performance capabilities of LSI composites.

When the IOWN era dawns, countless computing resources ranging from devices and edge servers to data centers will be connected by large-bandwidth optical communications. Post-Moore's Law technology is anticipated to boost computing processes by coordinating the functions of innumerable computing resources connected optically at high speeds.

Developed as one of the pioneering trials of this post-Moore's Law technology, optical interconnect was developed to increase the performance of information processing systems by amalgamating the high-speed protocol technology and communication processing circuit technology for optical communications that NTT has been

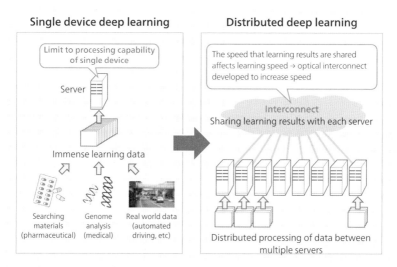

Single device deep learning

Limit to processing capability of single device

Server

Immense learning data

Searching materials (pharmaceutical)

Genome analysis (medical)

Real world data (automated driving, etc)

Distributed deep learning

The speed that learning results are shared affects learning speed → optical interconnect developed to increase speed

Interconnect
Sharing learning results with each server

Distributed processing of data between multiple servers

Fig.9-1

channeling its R&D efforts into.

Now we are putting AI in application, with the aim of developing faster *Distributed Deep Learning* that is capable of distributed processing immense volumes of data like automated driving or genome analysis spanning multiple servers. The training speed of AI depends largely on the time it takes to share the results of training between numerous servers **(Fig. 9-1)**. Increasing the speed that such results can be sent between servers was the next step.

Sharing the results of training at faster speeds

The first key point for achieving this concept is to build a closer link between the GPU that processes the training and the optical interconnect, with direct GPU-optical interconnect communications. With current computer technology, communications between servers inevitably result in data being exchanged between devices doing the computing within the server, and optical interconnect devices in charge of communications between servers. Creating a closer link between both ends and minimizing the speed for exchanging data between them greatly affects the communication speed. Rather than transmitting data via CPUs and main memories, the associated transmission delay can be slashed if the system is designed such that data can be exchanged directly between training processing GPUs and optical interconnect devices.

The second point was to develop a protocol and communications architecture suited to *collective communications* for sharing data between multiple servers. There is a better way than simply transmitting data from server A to server B when sharing data between multiple servers. The method dubbed *collective communications* uses data sharing concepts that combine data transmission with specific tasks like *summing up the computing results from multiple servers*, *distributing the same copy of results to all servers*, and *collecting data on all servers in a single location*. Increasing the speed of *collective communications* can boost the performance of multiple servers through cooperative computing.

The first step in the approach taken at this point was aiming to achieve faster speeds through the operation used for *collective communications* (Fig. 9-2) that takes up large part of processing for distributed deep learning: *collect all data residing on each server, sum them up, and distribute the summation results to all servers.*

More specifically, connecting each server to a process ring and adding to data while the data is sent to the adjacent server reduces the processing time for summing the data compared to collecting data from all servers and then summing them up. Additionally, summed up data can be distributed in the reverse route around the ring, allowing the processes for summing and distribution to be performed concurrently (Fig. 9-3). The communication protocol also comprises a simple structure to suit this

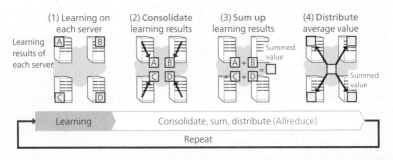

Fig.9-2

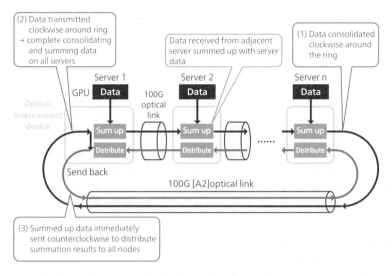

(2) Data transmitted clockwise around ring → complete consolidating and summing data on all servers

Data received from adjacent server summed up with server data

(1) Data consolidated clockwise around the ring

Server 1

GPU Data   100G optical link

Server 2   Data

Server n   Data

Optical interconnect device

Sum up

Sum up

Sum up

Distribute

Distribute

Distribute

Send back

100G [A2]optical link

(3) Summed up data immediately sent counterclockwise to distribute summation results to all nodes

Fig.9-3

operation to reduce protocol processing delays.

The third point involved developing an accelerator circuit as dedicated hardware to perform these data sharing processes at high speed. This design allows the summing and protocol processing, outlined in the second point above, to be performed at super fast speeds with collective communications by utilizing the accelerator circuit capable of processing at ultra-fast throughput on par with optical communications.

Core technology for future information processing systems

We ran a test to compare the performance of a system using this technology, against the fastest commercially available device that is in use today. [1] The results showed that with four servers connected (per GPU), the wait time for processing transfers (communication overhead) was reduced by more than 84% (Fig. 9-4), and the training speed increased by 7%. [2] Extrapolating from these observed results and increasing the number of GPUs indicates a 40% or more increase in training speed when 32 GPUs are used. [3]

This technology will soon be applied to data centers that perform AI training, with

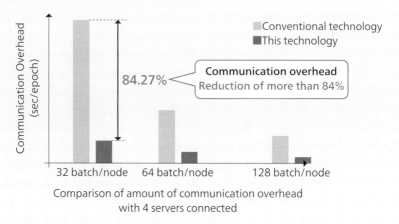

Comparison of amount of communication overhead
with 4 servers connected

Fig.9-4

the aim of amplifying the performance of AI training processes and reducing power consumption when handling vast amounts of data such as those associated with automated driving, genetic analysis or weather forecasts.

Looking forward, we will continue to find avenues where we can apply the technology we have developed to a broad range of fields other than AI. It is hoped that in the leadup to the post-Moore's Law era, this will form the core technology for future information processing systems that serve to boost performance by coordinating myriad resources connected to networks.

[1] Combination of 100 Gbit/s InifiBand + newest GPU available on the market
[2] Comparison tested with dataset: Imagenet, learning model: ResNet50, 4 servers, 1 GPU/server
[3] Estimate based on 4 servers, 8 GPUs/server

06 Networks

Achieving faster speeds, lower power consumption and lower latency in anticipation of post-5G requirements

There are many ideas being raised today on how to best utilize the 5G pre-service network that was rolled out in Japan in September 2019. Communication networks are constantly evolving as they are an essential component of social infrastructure. The global market for networks is also clearly growing in line with advances made to technology. People around the world are quite literally becoming connected via networks, and there will be an explosive growth in the number of IoT devices, connected cars and all things *connected*. The demand for networks will also skyrocket in line with the growing number of such devices.

So what exactly is going to be expected of these networks? The Ministry of Internal Affairs and Communications has announced that networks will need to meet the following six requirements by the year 2030: *Higher capacity*; *Lower power consumption*; *Ultra-low latency*; *Flexibility and high resilience*; *High-efficiency data distribution*; and *Safety and reliability* [1]. These six requirements are considered essential conditions for ensuring that networks become the real *infrastructure* underpinning our society. Numerous research institutions are taking two approaches for R&D that addresses the key challenges to fulfilling these requirements: *Faster networks*, and *More advanced network control.*

Adopting technological innovation to distribute data more efficiently at a lower cost

Faster networks literally refer to developing networks capable of sending and receiving huge amounts of data at high speeds and with low latency. It might be obvious that if there is an increase in network facilities, a high-speed network can be developed even with massive amounts of data. Yet this may not always be feasible due to the costs and materials involved, as well as the power consumption required. More advanced optical transmission technologies like space multiplexing or wavelength multiplexing are currently being developed around the world with the aim of creating faster networks and more compact equipment.

And while competition with development is heating up between corporations and research institutions, there are also cases of numerous companies working together to develop new networks. An emerging trend is where major tech companies like Google, Facebook and Microsoft are spearheading development of *communities* by applying the technical knowhow of numerous companies, rather than creating their own internal corporate networks.

On the other hand there are concerns that as network traffic continues to grow increasingly larger and more diverse, networks will not be able to handle the increase in communications requirements unless the functions of the network itself become more sophisticated. To address these concerns, organizations are attempting to develop adaptable networks capable of handling even more diverse requirements like edge computing, network virtualization, network optimization, automated operations using AI, and more advanced wireless technology. As the amount of data being transmitted continues to increase at a faster rate, data will need to be distributed more cost-effectively and efficiently. Research into automation of network control systems is likely ramp up in line with this.

Taking the initiative for innovative networks

NTT already operates an extensive span of network infrastructure, but still we are taking the initiative for developing new types of networks. NTT is developing cloud native networks that serve as social infrastructure capable of meeting the requirements for immense amounts of diverse resources, in line with Cognitive Foundation architecture that enables quick and efficient deployment of ICT resources. We are currently embarking of development of innovative wireless access/host IP architecture with the view of bringing sophisticated networks to market quicker. Just like research institutions in other countries, we are also pressing forward with further research into optical technology for transmitting huge amounts of data over longer distances. At the same time as we are focusing efforts on research of such elemental technologies, we are collaborating efforts with numerous partners to develop the groundwork for a *community* for next-generation networks.

In addition to research into next-generation access services or high-precision positioning technology, we are planning to launch a network feasibility study on the topic of MaaS as part of development of new networks operating as the channels of social infrastructure. High-precision positioning technology capable of accuracy on a centimeter scale will no doubt provide immense added value in the future for processing complex information in real-time via connected cars or IoT devices. Rather than viewing networks simply as a means of providing communications, it may be possible to identify a range of new services if networks are instead viewed as the core social infrastructure that smart worlds are built upon.

Shifting scales from *terabytes* to *petabytes*

As outlined previously, we are channeling our research efforts toward innovative networks built on cutting-edge optical technology, and are pushing forward with development of an extensive range of network technology. One example of this is network *intelligence*, where a network makes its own decisions and makes changes to its logical layout in order to achieve its goals, without human intervention. By fusing networks with AI, we may be giving those networks a way to alter their functionality in an autonomous manner.

We are also expanding our efforts for the development of ultra-large transport protocols that far exceed today's network capabilities, like new communication methods suited to edge computing, 1-terabyte class wireless and 1-petabyte class optical transmissions. Other efforts also include wavelength selector technology that enables flexible use of optical paths and fiber-optic environmental monitoring where optical fibers are used as new sensors for social infrastructure.

Alongside development of such elemental technologies and studies into new types of use cases, we are also turning our gaze to the development and maintenance of infrastructure software that utilizes open source software (OSS) ecosystems and integration verifications. While efforts are being rolled out for maintenance and improvements to infrastructure already in use—like converting existing networks to *all-photonics* through to individual terminals to create faster, more secure networks—new communities are also being linked into to infrastructure to give networks an additional boost.

Given the amount of work being made to develop a smart world, networking technology is taking on a much greater level of significance. Today, many people are anticipating the innovation that is slated to come with 5G networks. Yet we have set our sights so much further, as we have already embarked on projects to create even faster and more advanced networks. The innovative networks that NTT is working on will no doubt form the infrastructure required for a rich and intelligent society by bringing together market players spanning beyond national boundaries or industry sectors.

[1] Report from "Study Group on Future Network Infrastructure."

Case Study 10

Optical Fiber for Spatial Division Multiplexing
Backbone of the all-photonics network

Internet traffic is increasing at a rate of 30 to 50% annually, with the capacity of optical fibers said to reach their limiting capacity in the late 2020s. While technologies that *time division multiplexing or wavelength division multiplexing* are already being used, there is a growing trend of researching ways to further increase capacity by applying the new approach of *space division multiplexing* as a method to overcome these limitations.

Optical fibers are considered the arteries of the network in the all-photonics network at the heart of IOWN, and they form one of the extremely important core technologies behind the concept. We are leading the world in terms of research into such space-division multiplexing optical fibers, and apply it to push ahead with the IOWN design.

A 100-fold increase in cross-sectional area spatial multiplexing density

In the past, a single optical fiber traditionally consisted of a single light path (core) that allowed only a single-light (mode) to be transmitted. In contrast, space division multiplexing can be achieved by using multi-core optical fibers containing multiple cores within a single fiber, or multi-mode optical fibers capable of transmitting multiple forms of light (modes) within a single core.

The key challenge with multi-core optical fibers is preventing interference between optical signals transmitted along adjacent cores. We designed the core layout and distance between adjacent cores to maximize the spatial multiplexing density while retaining almost the same transmission characteristics as current optical fibers. The feasibility of core multiplexing exceeding 10 cores has already been verified.

With multi-mode optical fibers, the speed of the transmitted light varies with the mode, which increases signal processing complexity at the receiver end, and also sig-

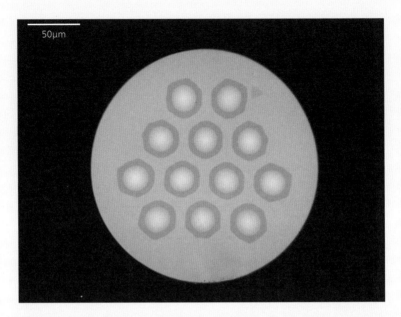

Fig.10-1

nificantly limits the distance that signals can be transmitted. To overcome these challenges, we have optimized the core structure (relative index difference in the radial direction) to design a multi-mode optical fiber capable of transmitting 10 types of modes along a single core.

The degree of spatial multiplexing can be increased significantly if we are able to develop a multi-mode, multi-core optical fiber combining multi-mode and multi-core technologies. We have applied the design guides we developed to successfully design and build a multi-mode, multi-core optical fiber with 120 (12 cores x 10 modes) spatial channels, the highest number in the world **(Fig. 10-1)**. This multi-mode, multi-core optical fiber has a cross-sectional area spatial multiplexing density more than 100-times that of existing optical fibers, which literally demonstrates a more than 100-fold increase in potential over current technology.

Delivering a petabit-class optical transmission system

The diameter of the multi-mode, multi-core optical fiber in Fig. 10-1 is approximately 220 μm, which is just under double the diameter (125 μm) of existing optical fibers. The manufacturing method of optical fibers involves making relatively large cylinders of glass (preform) which is then melted and stretched thin. For instance, a 10 cm diameter, 2 m long preform can be made into a 125 μm diameter optical fiber longer than 1000 km. Yet doubling the diameter of the optical fiber slashes the length it can be made to just a quarter.

We addressed this by examining ways to make multi-core optical fibers the same thickness as today's optical fibers. If we aimed for the same diameter as current fibers, the same manufacturing processes and connectors as existing optical fibers can be used. This would make developing this new type of optical fiber more economical and efficient.

We have already demonstrated the feasibility of running a 4-core fiber within the standard optical fiber diameter **(Fig. 10-2)**. This multi-core optical fiber was used as the world's first transmission line exceeding 300 km length successfully connected by multiple vendors. In another example, a 4-core optical fiber developed with low loss and minimal interference between adjacent cores was shown to be feasible as an undersea cable spanning some 10,000 km between two continents.

As we set our sights on developing petabit-class optical transmission systems, we will continue our R&D efforts into space division multiplexing optical fiber technology for its ease of manufacturing and efficient space multiplexing, and also work toward making this technology the industry standard.

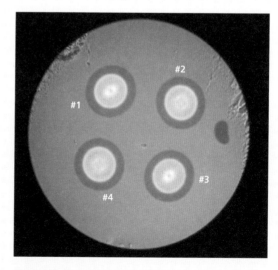

Fig.10-2

Case Study 11

Integrated optical front-end device technology
Enabling optical signal transmission for the All-Photonics Network (APN)

With the increasing uptake of new information and communication services like IoT and 5G, communications traffic is indeed likely to continue growing well into the future. There is growing demand to further boost capacity of such optical communication networks cost-effectively. To achieve this, developing more compact and higher density optical interfaces for transmitting optical signals, as well as increasing the transmission capacity per optical signal wavelength will be essential.

In order to satisfy such demands, NTT developed a compact Coherent Optical Sub Assembly (COSA) to operate as optical front-end integrated devices. Utilizing analog multiplexing, this technology was used to successfully transmit data over long distances at speeds of 1 terabit/s per wavelength, a ten-fold increase over conventional systems. These form the core technology powering the IOWN all-photonics transmission network.

World's smallest optical front end: COSA'

Silicon-photonics technology brings the high efficiency and dense integration of CMOS fabrication technology used in the production of large-scale integration (LSI) circuits to optical communication devices by applying the fine processing technology.. The technology allows optical interfaces to be more economical and more compact.

We were one of the first in the industry to put forward this technology as coherent optical sub-assemblies by developing optical front-end integrated devices based on digital coherent optical transmission technology.

Optical front-end functions required for digital coherent optical transmission comprise an optical modulator and driver for sending optical signals, and a coherent receiver and TIA (transimpedance amplifier). In the past, these functions were assembled by individual parts made with different materials and packages. All those parts had to be used by connecting them up with optical fibers or electrical cables.

The silicon-photonics technology that NTT has focused on developing is achieved with compact, high-density COSA, in which the optical modulator and coherent receiver are integrated onto a single silicon-photonics chip, with the TIA and drivers also implemented within the same package (Fig.11-1). This configuration consolidates the optical fibers, electrical connectors and individual parts that were required in the past, and achieves an economical device.

We worked on making COSA even more compact and faster using this silicon-photonics technology, and successfully designed COSA (Fig.11-2) to deliver optical signal speeds of 400 Gbit/s per single wavelength, with the smallest device in the world of its class. We used BGA (Ball Grid Array) package, a technology capable of economically mounting LSI, to make optical devices significantly more compact, and also applied NTT's proprietary optical circuit design technology to achieve the high level of reliability required of 400 Gbit/s-class optical communications infrastructure delivering the transmission speed of next-generation optical networks.

COSA can be applied to the smallest size of next-generation optical transceiver standards (QSFP-DD, OSFP). This technology is anticipated to be applied to medium- and short-haul transmission networks that will no doubt require more economical and higher capacity transmission signals in the future, then it leads up to all-photonics networks.

1 terabit/s long-haul transmission

In order to realize the high-capacity transport network for APN, it is necessary to achieve the high-speed long-haul optical channel transmission with the channel capacity over 1 Tbit/s per wavelength, by increasing the signal symbol rate [1] and available order number of multilevel modulation format.

For this purpose, it is indispensable to overcome the speed limit of silicon CMOS semiconductor circuits and to avoid signal degradation induced in ultra-high-speed signals interconnection between packaged devices. In order to overcome these limitations, we have proposed to combine novel optical digital signal processing and new device integration design using a new bandwidth doubler [2] based on analog multi-

Fig.11-1: COSA function block diagram based on silicon-photonics technology

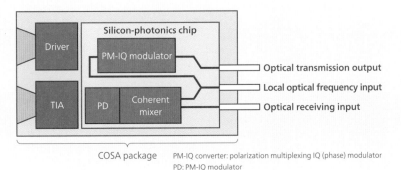

COSA package

PM-IQ converter: polarization multiplexing IQ (phase) modulator
PD: PM-IQ modulator
TIA: transimpedance amplifier

Fig.11-2: COSA module

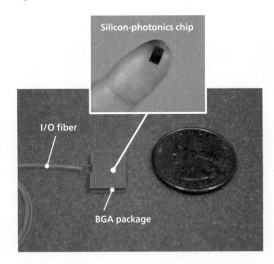

plexing (AMUX) [3]. Based on the new technologies, we have successfully generated and transmitted optical signals with symbol rate beyond 100 GBaud (transmitting data by switching the optical waveform 100 billion times per second).

For the long-haul transmission of the high-speed channels over 1 Tbit/s per wavelength, the generated optical signal quality has to be significantly enhanced. Therefore, we have designed elemental devices characteristics so as to maximize the total bandwidth of the integrated optical front-end circuit. And we also introduced the novel digital signal processing schemes that compensate the integrated optical front-end module imperfectness (such as differences in signal route length or losses caused in tributary signals) .

As results, we have successfully demonstrated the world's first long-haul wave-length division multiplexing transmission with the total capacity of 35 Tbit/s over 800 km, by applying NTT's proprietary digital signal processing technology and ultra-wide-band optical front-end integrated device technology **(Fig.11-3)**. Here we develop an inte-grated AMUX circuit chip based on our own Indium Phosphide (InP) HBT technology [4] **(Fig.11-4a)**, and achieved the channel capacity of 1 Tbit/s per wavelength to generate high-quality 120 GBaud PDM-PS-64QAM optical signals [5].

Furthermore, we have designed to develop compact and ultra-high-speed inte-grated optical front-end module that consists of two AMUX circuits and a novel wide-band InP optical modulator **(Fig.11-4b)**. The implementation of an analog multiplexing function (AMUX) into the optical front-end module makes it possible to reduce the mod-ule electrical input interface speed by the factor of 1/2, and the required module inter-connection bandwidth can be significantly relaxed in the electrical signal interconnec-tion between the silicon CMOS circuit and optical front-end circuit. As a result, we achieved a stable ultra-high-speed signal handling between circuit blocks.

The channel speed of 1 Tbit/s per wavelength that we demonstrated is 10-times higher than that in today's commercial system. These technologies expected to play a key role as the core technology that realize the innovative networks for future APN.

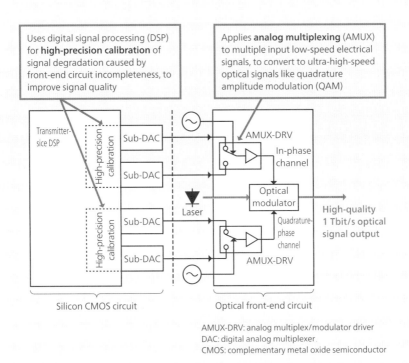

Uses digital signal processing (DSP) for **high-precision calibration** of signal degradation caused by front-end circuit incompleteness, to improve signal quality

Applies **analog multiplexing** (AMUX) to multiple input low-speed electrical signals, to convert to ultra-high-speed optical signals like quadrature amplitude modulation (QAM)

AMUX-DRV: analog multiplex/modulator driver
DAC: digital analog multiplexer
CMOS: complementary metal oxide semiconductor

Fig.11-3: Optical transmitter with built-in analog multiplex function (AMUX)

Fig.11-4 : Ultra-high-speed small-sized optical front-end module

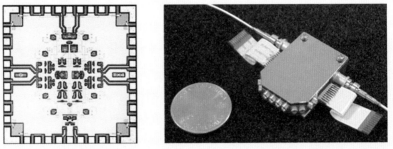

a   AMUX chip

b   Optical transmission front-end module with built-in InP optical modulator

[1] Number of times the optical waveform is switched per second. 100 GBaud optical signals transmit data by switching the optical waveform 100 billion times per second.

[2] Technology featuring a combination of digital signal processing, two DACs and AMUX to generate any signal double that of each DAC.

[3] Analog Multiplexer: an electrical circuit that switches and outputs alternately between two analog input signals based on a clock signal.

[4] Heterojunction bipolar transistor using an indium phosphide III-V group semiconductor. Capable of delivering superior speed and voltage resistance.

[5] Polarization Division Multiplexed Probabilistically Shaped 64QAM. This technology uses probabilistic distribution of optical signal fields into 64 individual signal point layouts to approach the theoretical data limit of signal error rates, to improve signal quality. This is used for each of the two distinct optical signals of polarization-division multiplexing (PDM) optical signals, which allows the transmission capacity to be further increased as a polarization-division multiplexing (PDM) optical signal.[5] Polarization Division Multiplexed Probabilistically Shaped 64QAM. This technology uses probabilistic distribution of optical signal fields into 64 individual signal point layouts to approach the theoretical data limit of signal error rates, to improve signal quality. This is used for each of the two distinct optical signals of polarization-division multiplexing (PDM) optical signals, which allows the transmission capacity to be further increased as a polarization-division multiplexing (PDM) optical signal.

07 Energy
Smart distribution of energy

The world's population continues to grow, and a United Nations survey predicts that it will exceed 10 billion people by the year 2060 [1]. Naturally, demand for energy will increase in line with this population growth. It is necessary to explore ways to use energy more efficiently, including the use of clean energy (renewable energy) developed with consideration for the global environment, and efforts to improve the balance between energy supply and demand, such as smart grid.

At the same time, the supply of energy is beginning to change. Households throughout Japan have traditionally purchased their energy from local power companies. In April 2016, with the liberalization of electricity retailing, households are now free to choose their power suppliers and services. And from April 2020, separation of the generation and transmission divisions of electric utilities (legal structural separation of power transmission and distribution) will take place. Consumers are already able to sell their unused solar power to electric utilities, and demonstration tests are already underway today to connect distributed power sources such as photovoltaic panels and battery storage (including electric cars), and trade energy among private users and companies. These are parts of efforts to encourage *local generation and local use* of power, or in other words, attain self-sufficiency.

There are worldwide movements today to create evaluation indicators for the environment, society, and governance as covered by ESG investments [2]. When choosing an electric power company, more people and companies are focusing not only on the price of electricity but also on whether they are using environmentally friendly renewable energy. With such movements underway, there will be an increase in available options in the future such as generating and storing power at home or at their own company and trading it with others without going through an electric utility (P2P energy trading), or buying clean energy that has been generated in an environmentally conscious manner.

Energy storage technology is the key to *local generation and local use of power*

For households or companies to generate or store their own power on the way to local generation and local use or self-sufficiency, advances in energy storage technology and new technologies to optimize energy distribution are needed.

One of the technologies drawing a lot of attention is high-capacity storage batteries. Of particular significance are the high-capacity, high-power output batteries installed in electric vehicles that are increasing in popularity. Batteries also play a vital role in developing smart cities or smart factories, as well as saving energy. They are also an essential component of BCP (business continuity plan) when the power supply grid is temporarily shut down due to disasters or similar incidents.

In the area of research related to storage batteries, the development of technology to replace conventional lithium-ion batteries is a particularly note-worthy trend toward *expansion of storage capacity*. As electrification is increasingly being used as the source of mobility, countries around the world are picking up the pace of research into novel battery alternatives, such as lithium-sulfur batteries and metal-air batteries that have a higher energy density than lithium-ion batteries.

Another research topic along with *expansion of storage capacity* is *ensuring safety*. Conventional lithium-ion batteries contain a combustible organic electrolyte. Further, as ions move between the positive and negative electrodes, metallic lithium is formed in a dendritic shape. These factors introduce the risks that can cause heat generation and explosions. Research and development of solid-state lithium-ion batteries that use flame-resistant solid electrolytes is being conducted on a national scale to prevent the formation of these dendrites and improve the safety of the batteries.

Flexible energy distribution with no wastage

In order to increase the utilization rate of green energy with consideration for the global environment, it is also necessary to optimize energy distribution without waste by creating and storing energy smartly. NTT is aiming to bring improvements to energy distribution by approaching the matter from three perspectives: establishing a *virtual energy distribution platform*; developing a *next-generation micro-grid*; and R&D into the underlying technologies that support these approaches.

The first, establishing a *virtual energy distribution platform*, is underpinned by virtual power plants. A virtual power plant collectively controls forms of energy storage such as storage batteries and electric cars owned by households and companies, and allows them to function as if they were a single *power plant*. We are developing technologies related to P2P energy trading utilizing block chain technology, as well as conducting our research on traceability to verify the path the energy takes before it reaches the end user. We are also pushing ahead with research to link up—or orchestrate—energy suppliers and consumers, and to meet energy supply and demand flexibly with high value-added. To achieve such goals, our development is focusing on large-scale, high-speed response virtual power plant technology that distributes energy *intelligently* among multiple virtual power plants, as well as real-time energy matching technology to respond to fluctuations in energy supply and demand.

Efficient distribution combining multiple energy sources

Our second area of focus is the *next-generation micro-grid*, which coordinates the various energy sources available within a region with the aim of distributing energy in a diverse and wide ranging manner. We are developing technologies aimed at optimizing costs by accommodating a hybrid energy system combining both alternating current and direct current power sources, as well as technologies needed to ensure that energy can be used safely and securely.

In addition to these two approaches, we are also channeling our R&D into various core technologies. Energy is likely to be delivered via various means in the future, and development of underlying technologies will lead to new distribution methods or applications for energy. When considering the use of light to deliver energy for instance, if the minute amount of energy transmitted via optical fibers could be converted to electricity efficiently, equipment that traditionally operated on electricity could be powered by optical energy as a backup solution.

Other development projects that we pursuing include core technologies related to clean energy, such as the Return to the earth battery (CS18) and artificial photosynthesis. Making environmentally conscious clean energy distribution around the world, a reality will require development of technologies to enable *local generation and local use* of power, and to distribute energy in the most optimum way possible. The smart world afforded by IOWN will also make the use of energy more intelligent. As the global population reaches 10 billion with an average lifespan approaching 100 years, we are committed to conducting continuous R&D so that no one suffers from a lack of energy, while at the same time helping to preserve the environment.

[1] World Population Prospects 2019
[2] Refers to investments with a high priority on the Environment, Social and Governance aspects.

Artificial photosynthesis technology
The technology of dreams to achieve a carbon-recycling society

One of the hottest topics of the 21st century is the environment and energy. Efforts to *prevent global warming and break away from fossil fuels* represent an important paradigm shift where *technology* and *nature* coexist together in harmony, instead of competing against each other.

One of the technologies that NTT anticipates as vital to prevent global warming and break away from fossil fuels is artificial photosynthesis technology. We are advancing research and development in this field with the aim of producing green fuels like hydrogen and methane from water and carbon dioxide by using light from the sun, just like plants do with photosynthesis. This is considered one of the key technologies for shifting to a carbon-recycling society, which seeks to prevent global warming by producing fuels derived from carbon dioxide **(Fig. 12-1)**.

Artificial photosynthesis to revolutionize lifestyle and society

People will be able to live a more environmentally friendly lifestyle if we are able to develop panels equipped with artificial photosynthesis technology, similar to the solar panels installed on the roofs of houses throughout our cities. Perhaps panels can be mounted on the top of bus stops, so that cars and buses powered by fuel cells can fill up with environmentally friendly fuel every time they come to a stop, helping to bring a greener solution to transportation. Development of artificial photosynthesis technology will ensure a lifestyle well balanced with nature, full with the benefits afforded by the sun. Looking further into the future, there may even be vending machines that sell environmentally friendly fuel, or food trucks adorned with the catchphrase: "Cooking with green fuel" **(Fig. 12-2)**.

Venture further out the city, and there could be large-scale solar panels dubbed mega-solar working in tandem with artificial photosynthesis panels to produce electricity and fuel as a reliable energy source for the local region. These areas would be

Carbon-recycling society

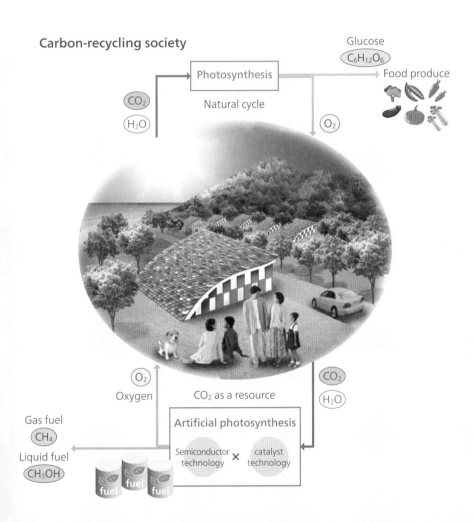

Fig.12-1: Society using fuels derived from CO2 to prevent global warming

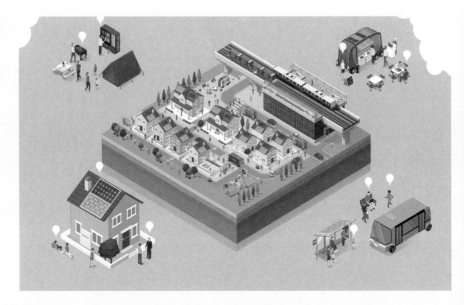

Fig.12-2: How artificial photosynthesis can change our lifestyle

self-sufficient as they generate and use energy within their local region.

NTT embarked on the first step toward a future like this in 2015 by kicking off research into artificial photosynthesis technology based on semiconductor growth technology and catalyst technology developed with R&D into data communications.

Artificial photosynthesis makes use of the following reaction processes. Electrons (e-) and holes (h+) are generated when semiconductor electrodes are exposed to sunlight. The holes (h+) utilize the oxidizing reaction of water (H_2O) to form oxygen (O_2) and protons (H+), while on the opposing metal electrode the electrons (e-) and protons (H+) induce a reduction reaction to form hydrogen (H_2). If carbon dioxide (CO_2) is also supplied to the metal electrode at the same time, products like carbon monoxide (CO), formic acid (HCOOH), methanol (CH_3OH) and methane (CH_4) are generated **(Fig. 12-3)**.

Research institutions like universities as well as companies are currently conducting basic research into artificial photosynthesis. The current stage of research is mainly focusing on determining the feasibility of practical applications, rather than

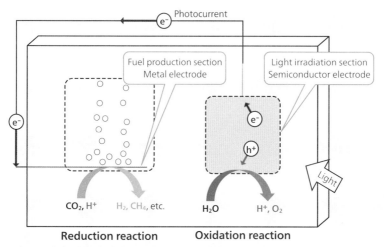

Reduction reaction   Oxidation reaction

- Electrons (e⁻) and holes (h⁺) are generated when semiconductor electrodes are irradiated by light
- Electrons on the metal electrode undergo reduction reaction to form fuel

Fig.12-3: Illustration of the artificial photosynthesis reaction process

studying the practical applications themselves. We have identified a number of issues during the course of our research efforts, with the main challenge being to achieve a balance between efficiency and operating life.

Tackling the challenge of greater efficiency and longer operating life

Efficiency here refers to how efficient hydrogen and oxygen can be produced from water using energy from the sun. Any practical applications will require achieving a major boost in efficiency compared to the photosynthesis used by plants (around 0.2 ~ 1.0%). In 2017, we were able to raise this efficiency to 0.84%. While the energy we produced with minuscule, the fact that we were able to successfully produce fuel more efficiently than plants is a major breakthrough in itself. With today's technology, we would need to reach an efficiency of 10% to bring costs down below that of fossil fuels. To achieve this, we are channeling our research into the materials and structures of semiconductor electrodes that are able to efficiently absorb visible light—the wave-

Known problems (semiconductor degradation due to corrosion reaction)

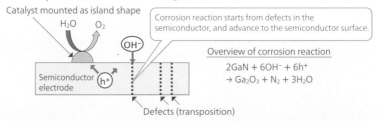

Catalyst mounted as island shape

Corrosion reaction starts from defects in the semiconductor, and advance to the semiconductor surface.

Overview of corrosion reaction

$$2GaN + 6OH^- + 6h^+$$
$$\rightarrow Ga_2O_3 + N_2 + 3H_2O$$

Semiconductor electrode

Defects (transposition)

Proposed solution (form catalyst protective layer and use low-defect semiconductors)

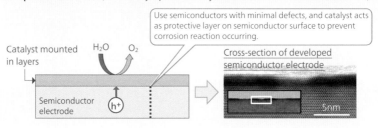

Use semiconductors with minimal defects, and catalyst acts as protective layer on semiconductor surface to prevent corrosion reaction occurring.

Catalyst mounted in layers

Cross-section of developed semiconductor electrode

Semiconductor electrode

5nm

Fig.12-4: Ideas to stop degradation in artificial photosynthesis

length that makes up the majority of energy from the sun.

More than just efficiency, operating life is also another vital factor that needs to be considered. Our studies have shown that when semiconductor electrodes are exposed to light, the semiconductors begin to corrode very quickly and efficiency drops off within hours. The cause of such a short operating life is that the holes (h+) formed when irradiated with light is actually the semiconductor itself corroding away **(Fig. 12-4 top)**.

We are trying to overcome these issues related to corrosion with two main approaches. The first is mounting the catalyst as a layer on the surface of the semiconductor. In this idea, the catalyst acts as protective layer to prevent the corrosion reaction from occurring **(Fig. 12-4 bottom)**. As the semiconductor electrode needs to be irradiated efficiently by light, initial designs had the catalyst placed in island shaped patterns to prevent them blocking the light. After further investigating the corrosion mechanism, we thought it would be a good idea to cover the entire surface with the catalyst. We kept the catalyst layer to around 1 to 2 nm thick (1/100000th the thickness of a human hair) to ensure that there is sufficient light transmittance.

We discovered that the corrosion starts from transposition (defects) in the semi-conductor, so the second approach was to use semiconductor electrodes with minimal transposition density (defects). As a result of this research, in 2017 we were at the level of 86% of the starting efficiency (photocurrent) after 100 hours, and by 2019 we had attained a level of 89% of the starting efficiency after 300 hours [1]. To put this figure into perspective, when exposed to six hours of sunlight per day, this equates to rooftop artificial photosynthesis reactions that can run continuously for more than a month and half.

Until 2017 we were focusing on hydrogen, which is a relatively easy fuel to pro-duce, however from 2018 we began turning our efforts to capturing carbon dioxide (turning it into fuel like methane) that will be essential for developing a carbon-recycling society.

Research into artificial photosynthesis is still in the fundamental stages, and there is still a long way to go before any practical applications will take shape. Yet it is important to remember that the Apollo program depicted in the film *Moonshot* [2] was a mission of immense proportions, requiring an unbelievable level of technology that redefined all common knowledge of the time. In a similar vein, we will push forward with our research and development efforts, applying the lessons learned by those before us. Yet instead of simply continuing where others left off, we will come up with innovative new ideas that may appear to defy common knowledge, and tackle issues face on to bring about major breakthroughs.

[1] This figure has been calibrated from the drop in performance of a light source lamp simulating sunlight.
[2] Refers to an immense undertaking or goal that, while extremely difficult as well as creative, will lead to innovation that will make a major impact.

08 Quantum Computing

From solving optimization problems, to revolutionizing the *concept of information processing*

Quantum computers are said to have incredible computing power that far surpasses conventional computers. Indeed, they are anticipated to be the key to solving problems that remain unsolvable to this day. This technology is thought to have enormous potential in a broad range of industries, and is today the focus of much attention by many research institutions and IT companies the world over. And while the technology is still in its early stages, the market size is predicted to explode at a high rate of 35% per year through to 2021 [1].

While the *quantum circuit model* extension of traditional classic computers is given as the model of quantum computation, more recently a completely different model has been proposed—the *Ising model*. The quantum circuit model is capable of running various quantum algorithms, and applications also include quantum simulations and security technology based on quantum circuit. In contrast, the Ising model is a quantum computer specializing in solving combinatorial optimization problems (acquiring approximate solutions) by converting those problems to mutually interacting quantum magnets, and using the property of physical phenomena where the energy naturally tends to the lowest state.

There is one application of quantum computing that is deemed to be available in a practical sense quite early, and that is *optimization problems* that are found throughout various industries. Using conventional computers for such an application requires an immense volume of calculations and takes an extremely long time to complete, and results may not even be attainable in some cases. Quantum computers is one solution for solving such problems. Examples where quantum computing are thought to be ideal for immediately returning the optimum answer from an immense number of options include: optimizing logistics, inventories or store locations; optimizing routes for mobility or population flows; and optimizing the locations to place city infrastructure.

Naturally, resolving optimization problems is just one example of the vast potential of quantum computing. As processing speed, processing accuracy and computing scale increase in size, the technology can also be applied to high-precision molecular simulations, which is expected to play a vital role in fields such as energy and drug discovery.

Two challenges that quantum bits are facing

There are two challenges that are drawing attention in the current field of research into quantum computing: *stabilizing quantum bits*; and *increasing the number of quantum bits*. That is, the instability of quantum bits and their insufficient number has posed a bottleneck for today's quantum computing. Overcoming these roadblocks will be essential in order to make positive progress with quantum computing.

The main problem with *stabilization* is that errors occur quite easily with quantum bits, as they are affected by noise from various factors during computations which results in altered quantum superpositions. This means that quantum bits can only preserve their information for a short period of time, and thus limits the time available for running computations.

Quantum error correction is one known method for overcoming these errors. Yet conventional quantum error correction algorithms are subject to physical limitations that make them difficult to implement in reality. A quantum error correction algorithm called surface codes was published by Eric Dennis et al. in 2002, where quantum bits can be arranged in a two-dimensional array. This method is drawing increased attention as the results can be applied to stabilize actual quantum bits [2].

The other challenge—*increasing the number of quantum bits*—comes about because the low quantity of quantum bits at the moment mean they can only be used for solving a limited number of problems. Bringing quantum computing to more industries in the future calls for increasing the number of quantum bits as well as raising their level of accuracy, and eventually achieving a high scale of integration on chips.

Numerous universities and research institutions are continuing their efforts to address these challenges, an example of which is the paper published by Dannna Rosenberg et al. at Massachusetts Institute of Technology (MIT) in 2017 covering the 3D integration of quantum chips [3]. 3D architecture is considered one method that is available for expanding the potential of quantum bits, and this paper examined how 3D architecture technology used for semiconductors could be applied to the field of quantum computing. Many research institutions are exploring a wide array of methods that can be used to *increase the number*, and the world is expecting a breakthrough in the near future.

Revolutionizing the concept of information processing

NTT has been continuously researching quantum information processing technology for more than three decades, a long time before quantum computing started drawing the level of attention it does today. There has been much progress in research aimed at developing circuit-model quantum computers. Yet there are numerous challenges in developing circuit-model quantum computers that need to provide *accuracy* and *universality*, and any realistic practical applications are still a long way off. On the other hand, while Ising model computers that use natural physical phenomena may find it difficult to come up with exact solutions, these computers are readily scalable and can provide practical solutions quickly. Using new computational methods or architecture based on ideas from the Ising model is thought to be a way of approaching practical combinatorial optimization problems, and practical applications based on new forms of information processing technology are likely to become a reality.

Yet the technology behind quantum information processing has more potential than merely software applications like resolving optimization problems or encryption processing. Research is also focusing on the hardware side, like new high-performance devices made with superconducting quantum bits and new devices featuring topological insulators, as well as the algorithms needed to operate them. Just like Ising model computers are doing, research into these fields is leading to a new phase by taking a different approach to existing quantum computing research.

Quantum computing is often viewed as a means of solving problems that existing computers were unable to. Yet beyond the constraints of existing quantum computing, these new computing methods and architectures show that quantum computing is leading to a *new concept of information processing*. For instance, optimization problems that take time to find exact solutions can instead be addressed with a new approach to information processing, where an approximate answer can be found quickly by using the analog properties of physical phenomena or new Ising model computations. Indeed, this could become the new way of processing information in the smart world.

[1] Technavio, Global Quantum Computing Market 2017-2021

[2] Paper published in "Journal of Mathematical Physics" Vol. 43 (Eric Dennis et al., "Topological quantum memory")

[3] Paper published in "npj Quantum Information" Vol. 3 (D. Rosenberg et al., "3D integrated superconducting qubits")

Case Study 13

LASOLV
Revolutionizing computations for combinatorial optimization problems

Combinatorial optimization problems—where the optimum combination needs to be found from a vast number of options—are known as difficult problems to solve for existing digital computers. More recently, combinatorial optimization problems are being converted to the *Ising model, which is a theoretical model to describe* mutually interacting spins. There is much research being conducted using this to solve various optimization problems quickly by determining the spin arrangement with the lowest energy using physical experiments using artificial spins. NTT developed a *coherent Ising machine* (CIM) that solves Ising model problems using an optical oscillator called an optical parametric oscillator (OPO).

Tackling combinatorial optimization problems with light

Spin is the concept of quantizing the angular momentum of particles as either a *spin up / spin down* state, or a superposition of these states. The Ising model is a theoretical model that describes the behavior of interacting spins, for which NTT uses OPOs to simulate. An OPO is a laser oscillator that uses a phase sensitive amplifier (PSA). Since the PSA amplifies the 0 and the π phase components of input light most efficiently, an OPO oscillates only at the phase 0 or π. By assigning the 0 / π phase of the OPO to a spin value +1 (spin up) / -1 (spin down) , we can represent the state of a spin reliably using a high-frequency optical oscillator.

We generate several thousand time-multiplexed OPO pulses by turning on and off a PSA placed in a 1-km optical fiber ring at a high-repetition frequency (typically 1 GHz). We use a scheme called measurement and feedback to implement mutual interactions among the OPO pulses. With this scheme, the amplitudes of all pulses are measured by extracting part of the pulse energies with a beam splitter for each circulation of OPO pulses within the ring. A feedback signal is then calculated with an electronic circuit for each OPO pulse based on the measurement results. This signal is convoluted to a light pulse, which is injected into the OPO pulse in the ring to complete OPO interactions. Since the OPO network tends to oscillate at a phase configuration with the lowest energy, the ground state of the given Ising model can be obtained quickly and with a high probability.

Unveiling a new realm of information processing

NTT has already developed a CIM capable of all-to-all interactions between 2000 OPO pulses, with which we demonstrated that the solution to a Max-Cut problem [1] of up to 2000 nodes can be obtained quickly.

A paper published in the American journal "Science" in 2016 reported that CIM was capable of solving problems to a similar level of accuracy an order of magnitude faster than *simulated annealing* [2] implemented on a CPU. In 2019, it was confirmed that the CIM obtained the exact solutions for Max-Cut of dense graphs with much higher probabilities than a quantum annealing machine based on superconducting devices. We also developed the *LASOLV* CIM system that features enhanced optical stability in a compact housing. With this system, we are now investigating potential CIM applications together with various partners.

The CIM represents a new form of computing developed through the culmination of core technologies being developed at NTT, including quantum optics, optical device and optical fiber system technologies. As stated above, the CIM has already shown the potential of outperforming existing digital computers or quantum annealing devices for tackling certain problems. Yet there are still so many questions that remain unanswered. For example, what role the analogness, nonlinearity, quantumness, and other aspects of OPO will play in the CIM computation, as well as finding out where the origin of the CIM computing power exists. By addressing these challenges, we are hoping that CIM will become a computational system that will play a role in overcoming the *saturation of Moore's Law*.

[1] The problem of determining how to divide nodes in a graph into two groups so that the number of edges between the nodes in different groups is maximized

[2] Simulated annealing is a method that uses the analogy of annealing used when machining steel. It is one method of finding an approximate solution to an optimization problem, by using temperature figuratively as a parameter in a probabilistic search when changing temperatures from high to low. Simulated annealing is a heuristic algorithm used to find many approximate optimums of a lower accuracy, when solving for global optimization in a large search space.

Case Study 14

Proof-of-principle experiment for quantum repeaters using all-photonic quantum repeater scheme
Paving the way for a quantum internet

NTT worked with Osaka University, University of Toyama, and University of Toronto to demonstrate the world's first successful proof-of-principle experiment toward developing *all-photonic quantum repeaters*. This is a scheme that is capable of achieving a worldwide quantum network comprising only optical devices. Quantum repeaters are essential to develop a *quantum internet* that would be considered the absolute ultimate information processing network. Such a network would have immense new potential that would go far beyond that of the internet we know today. They are also closely related to the concept of an *all-photonic network*, in which all the communications used for today's internet would be replaced with optical devices. This is an extremely promising vision that aims to develop a high-speed, low-power consumption network. Our successful proof-of-principle experiment can be viewed as our first major step as humanity toward developing a worldwide all-photonic quantum internet.

The ultimate information processing network

"Information is physical." This famous quote by Rolf Landauer (1927 to 1999) of IBM posited that information is always expressed by a physical system. Since the physical system obeys the laws of physics, the limits of information processing should follow the same laws. In modern physics, the most elaborate description of the natural world is afforded by quantum mechanics. The computer predicted within the framework of quantum mechanics is the quantum computer, while the allowed communication is quantum communication. In other words, according to the proverbial words of Landauer, *quantum information processing* comprising these quantum computers and quantum communications is the ultimate form of information processing as permitted by the laws of physics.

Given that today's internet is regarded as the largest computer network in the world linking information terminals of all clients on the planet, the *quantum internet* is the quantum version of this, playing a role in linking the *quantum* information terminals of any client on earth to create what could be the ultimate information processing network as permitted under quantum mechanics.

It is in fact known that a quantum internet like this could be capable delivering innovative functions beyond the framework of the internet as it is today. An example to highlight this is to provide quantum cryptographic communication to arbitrary users on the network to protect against any hackings (including ones based on a quantum computer). This cryptography has the highest level of security, and would allow data to be transmitted for purposes such as national referendums, summit meetings, financial transactions, genetic information, and even biometric information. There is no doubt that the technology will be used to create new forms of electronic money.

The quantum internet could transfer unknown quantum information faithfully to people located remotely through the use of quantum teleportation. This will form the foundation of distributed quantum computation, cloud quantum computing or for building quantum computer networks.

The quantum internet could also be used in synchronization with atomic clocks—the most accurate clocks existing today—which would allow time to be coordinated on a worldwide scale with the most secure, accurate and safest timekeeper available. Applications anticipated to harness this potential would include high-precision navigation systems. It would also play a role in advancing astronomy by extending the baselines of telescope arrays.

Quantum repeaters using only optical devices

The potential applications of quantum internet span far and wide. In order to use existing optical fibers already available around the world, the repeaters that are currently installed to amplify the light needs to be switched over to *quantum* repeaters. To achieve these quantum repeaters, a matter quantum memory [1] was previously deemed essential. Going against this dogma, however, we collaborated with the University of Toronto in 2015 to propose the all-photonic quantum repeater scheme as a way developing the required quantum repeaters using only optical devices, without the need for matter quantum memories at all.

If we are able to develop an *all-photonic* quantum network using this scheme, it would deliver a host of benefits not available with networks based on matter quantum

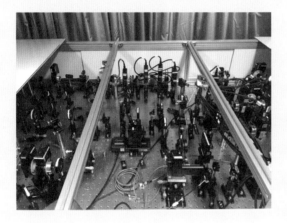

Fig.14-1

memories. A few examples include its *communication rates independent of communication distances to achieve a high-speed quantum internet*; *no interface needed between photons and matter, so it is energy-efficient and operates at room temperature*; and *easier to achieve than a similar (all-optical) quantum computer, representing a major milestone in the development of quantum computers*.

In 2019, we worked together with Osaka University, Toyama University and the University of Toronto to run the world's first experimental demonstration of the *time-reversal* idea that is at the core of these all-photonics quantum repeaters **(Fig. 14-1)**. In this experiment based on the concept of time reversal, quantum information carried by a photon was teleported to a different photon located separately without disturbance from photon loss, as the theory predicts.

With the theory behind these all-photonics quantum repeaters having been demonstrated, further progress with research and development into optical devices like low-loss integrated optical circuits or efficient photon sources is anticipated. If this can be achieved, we may one day be able to create an all-photonic quantum network built with all-photonic quantum repeaters, as well as an all-photonics quantum computer. Then, developing an all-photonic quantum internet based on these technologies may not be a dream after all.

[1] In addition to 0s and 1s, this memory can also store each quantum mechanical superposition state. Matter quantum memory is a quantum memory developed with actual matter.

Case Study 15

Nanophotonic device technology
Underpinning the future IOWN platform

Developing "optical computers" had long been one of the goals of researchers in the field of photonics, but with LSI electronics currently at the peak of their integrated performance, research into optical computers has waned as they were no longer deemed significant. But recently, with the limit to the level of LSI miniaturization and integration approaching, there is growing demand that optical technology can be used for information processing in cooperation with electronic circuits, instead of simply long-haul signal transmission in optical fibers. Providing greater momentum to such demand is the fact that light is being re-evaluated in recent years for its potential not only for digital processing but also for machine learning and other forms of analog processing. Accordingly, optical information processers linking both electronics and optics are gradually beginning to take shape.

The key to unleashing this potential is opto-electronic conversion technology. NTT has successfully used nanophotonic technology to drastically reduce opto-electronic conversion energy loss, and applied the technology in a demonstration of unprecedented optical transistor. These results represent a major milestone for information processing that closely integrates optical and electronic circuits, and is demanded to be utilized for the hardware technology underpinning the future IOWN framework.

Photonic crystals to achieve a breakthrough in opto-electronics

One of the challenges with innovative information processing technology is achieving more compact and energy-saving devices that convert between optical and electrical signals as a way of developing a dense photonics-LSI interface. Opto-electronic converters in the past had a very large capacitance, and operating them required high energy consumption some 100-times that of the electronic transistors used in LSIs, which caused a major bottleneck.

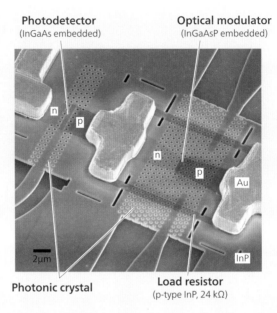

Photodetector
(InGaAs embedded)

Optical modulator
(InGaAsP embedded)

Photonic crystal

Load resistor
(p-type InP, 24 kΩ)

Fig.15-1: Optical transistor fabricated on photonic crystal platform. Photo viewed with scanning electron microscope.

NTT has worked to resolve this issue by using artificial nanostructures called photonic crystals. Photonic crystals are structures with a periodically modulated refractive index on a sub-micron scale that are capable of trapping light in tiny areas.

This research uses a photonic crystal with periodic air holes formed in a thin indium phosphide (InP) semiconductor plate **(Fig. 15-1)**. If the areas without holes are arranged carefully, light can be trapped in the specific regions to form tiny optical waveguides or optical resonators. Photonic crystal devices are being studied in earnest around the world. However, NTT has developed a distinct fabrication technique dubbed the buried heterostructure, capable of forming functional materials within these tiny optical waveguides/resonators with a high level of precision. This technology enables fabrication of a broad range of photonic devices, and NTT has demonstrated record-low energy consumption of optical switches, optical memories, laser sources and other photonic devices. As outlined below, this technology has also successfully been used to develop opto-electronic converters that are smaller in size, with a lower capacitance and lower energy consumption.

Developing ultralow energy consumption opto-electronic converters and optical transistors

To fabricate a photodetector capable of converting optical signals to electrical signals (O-E conversion), an indium gallium arsenide (InGaAs) absorption material was embedded into the optical waveguide. On the other hand, to fabricate an electro-optic modulator capable of converting electrical signals to optical signals (E-O conversion), an indium gallium arsenide phosphide (InGaAsP) with strong optical nonlinear effects was embedded into the optical resonator. Both of these devices are on a scale of several μm, which is extremely compact compared to existing photodetectors and modulators. Their capacitance can be minimized to the same level as transistors inside LSIs. NTT applies this ultralow-capacitance technology for demonstrating the potential of photodetectors with no electrical energy consumption, and also achieved optical modulator with lowest energy consumption in the world. This breakthrough in O-E/E-O converters suggests the potential for seamless opto-electronic interfaces within information processing chips.

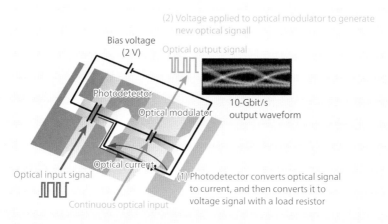

Fig.15-2: Operation principle of optical transistor operation

NTT also demonstrated an optical-electrical-optical (O-E-O) converter with the photodetector and optical modulator integrated in close proximity as proof of ultra-low-energy consumption opto-electronic processing **(Fig. 15-2)**. Optical signals input to the photodetector are converted to current, and then converted to a voltage signal by a high 24 kΩ load resistor. This voltage signal directly drives the optical modulator to output a new optical signal with a signal gain. This resulted in a nonlinear optical-to-optical signal conversion with a bitrate of 10 Gbit/s. This operates with an optical input energy of several femtojoules (fJ) per bit, which is around the same level of energy an LSI transistors, making it the first demonstration of opto-electronic operation of this level in the world. This integration comprises an optical input to the photodetector, optical input to and output from the optical modulator. With such a three-terminal optical device, in which the output signal power from the optical modulator can be larger than the input signal power to the photodetector, NTT developed an optical transistor just like LSI transistors.

This low-capacitance opto-electronic device has also resulted in energy consumption being reduced to 1/100th that of existing three-terminal optical devices. Having the optical signal gain means that the optical signal can be transferred in multiple stages, with a broad range of applications expected in the future.

One application to illustrate this is development of an integrated photonic network chip using optical transistors as repeaters to share cache data between many-core CPUs. Other examples include optical neural networks and machine learning, where optical transistors may function as nonlinear devices or neurons in the optical domain. These technologies are likely to drive the shift in our mindset from separative "optical transmission and electrical processing" to more collaborative "opto-electronic information processing".

09 Biomedical

From molecular level design, to research spanning multiple fields

When people hear the term biotechnology, they may first imagine modified foods or varieties of agricultural produce (like genetic recombination) that were hot topics in the past. These days, the term may bring to mind new developments in medical treatment like CRISPR genome editing technology. Yet today, biotechnology is used for applications in a broad range of industries beyond that of medical or agriculture, forestry and fisheries.

For example, bio-identification technology is useful when verifying identities for financial transactions, and of course biometric identification is widely used for smartphone authentication these days. Greater biomass power generation efficiency could also help resolve challenges related to energy. And in manufacturing, nano-level molecular manipulation could unlock the potential for designs starting from the base materials, with biochips and biomarkers already being utilized for practical purposes. Biotechnology is indeed a technology that will be closely intertwined with the telecommunications sector.

High-precision, multidimensional designs

Today there are two topics that are drawing the focus of research efforts in the realm of biotechnology: molecular scale manipulation, and multidimensional molecular design. Biotechnology allows operations to be performed on materials on a much smaller scale, down to the molecular level. Molecular scale manipulation increases the precision of such operations, and can be used to induce desired properties or functions within the body. To highlight an example, a paper published in 2018 by Long-Hai Wang et al. at the Chinese Academy of Sciences was on research focusing on efficient methods for gene insertion. The paper demonstrated how cationic polymers can be used as a vector for gene transfection when inserting a particular gene's DNA for gene expression or recombinant protein preparation [1]. As research efforts like this continue to advance, we will be able to more accurately induce the properties we are targeting.

Until now, biotechnology had been used to manipulate cells and molecules in order to induce changes, but performing manipulations that take time had proven extremely difficult, like ensuring changes occur before a specific time. One way over overcoming this that is drawing increased attention is multidimensional molecular design.

Trends in multidimensional molecular design can be seen in a paper published in 2016 by Alec Nielsen et al. at the Massachusetts Institute of Technology and researchers at the National Institute of Standards and Technology (NIST) and other institutions [2]. The paper outlines how the reactions (synthesis or decomposition) of cells to their environment can be utilized to program cells to undergo specific reactions. If such programming can be applied on an individual cell level, it may lead to potential developments in the future where we can control the autonomous component generation of cells as well as hone the timing of changes and amount generated.

And when it comes to high-precision and multidimensional design on a cell level, regenerative medicine with the use of iPS cells can be considered one of the key topics. In 2019, a team led by Professor Koji Nishida of Osaka University were the first in the world to transplant corneal tissue sheets derived from iPS cells onto a cornea. This technology will play a role in restoring the eyesight of many patients suffering from loss of vision due to corneal disease, for which treatment had been difficult. Clinical studies involving iPS cell transplants are already underway for patients suffering from corneal disease, heart failure, spinal cord injury, Parkinson's disease and other illnesses, and the research may hold the key to developing the ultimate treatment methods leading to complete cures.

Harnessing our knowledge of biosensing and biomedical

NTT is also turning its gaze to challenges like these as it works to expand research across multiple fields. This broad approach to research covers many areas that are advancing the field of medicine, including: biomimetic nanodevices and biosensing that incorporate biological systems, AI analysis of personal medical data, biomonitoring throughout daily life, and biocompatible material to supplement damaged biological functions.

One field we are focusing on in particular is the area of biomedical care. In 2019 we teamed up with Australian universities Deakin University and Western Sydney University to embark on research for achieving the vision of a society where the elderly can live healthy, independent, and safe lives. We have launched a number of joint research projects with the aim of starting demonstration experiments first in Australia from 2020. Projects include research and development to support communication between dementia patients and their families and caregivers, and smart homes designed that have been designed for the elderly and disabled to live safely and securely.

Elderly care is one of the fields of medicine that will be important in Japan in the future. With this in mind, we are harnessing our knowledge in the fields of biosensing and biomedical care to push forward research beyond our own country.

Further advances to multidisciplinary research in the biomedical field helps to not only improve QOL (quality of life) of people and the environment, but also forms the foundation for supporting the development of a smart and affluent society.

[1] Paper published in "Angewandte Chemie" Vol. 55 (Long-Hai Wang et al., "High DNA-Binding Affinity and Gene-Transfection Efficacy of Bioreducible Cationic Nanomicelles with a Fluorinated Core").
[2] Paper published in "Science" Vol. 352 (Alec A. K. Nielsen et al., "Genetic circuit design automation").

Case Study 16

hitoe®, performance material for collecting biosignals
A better understanding of mind and body conditions just by wearing *hitoe*

Living a rich, healthy life requires early detection and treatment of illnesses. Daily monitoring of mind and body conditions is the most effective method for achieving this. To reach this goal, NTT DOCOMO and Goldwin teamed up together in 2014 to develop innerwear capable of measuring biosignals. This innerwear measures the user's heart rate and ECG signals when it is simply worn, and is chiefly used for managing sports training and enhancing performance. The innerwear includes bioelectrodes which are made of a functional material called *hitoe*® that was jointly developed by Toray Industries and NTT. The material features ultra-fine fibers and offers excellent moisture retention and affinity with the skin—the bioelectrodes of the human body. More than just for sports applications, *hitoe*® is expected to be used for monitoring the body's physical condition.

Smart clothing for monitoring body and mind

hitoe is made by firmly coating the advanced nanofiber fabric developed by Toray with conductive polymer PEDOT:PSS [poly (3,4-ethylenedioxythiophene)-poly (styrenesulfonate)] to create a fabric bioelectrode that is gentle on the skin. It offers superior flexibility, stretchability, breathability and bio-affinity, and differs to traditional medical electrodes in that it can measure heart rate, ECG, EMG and other biosignals with a high level of precision without using metal electrodes or electrolyte paste. The nanofiber making up the base fabric is an ultrafine material that is extremely thin with a diameter of around 700 nm. This nanofiber helps to achieve a tighter fit with the skin, which enables biosignals to be measured reliably with low noise.

hitoe electrodes can also be combined with inner shirts to develop a wearable system capable of measuring the wearer's heart rate and ECG simply by wearing the clothes. In 2014, Goldwin and NTT DOCOMO released the sports innerwear C3fit IN-pulse for measuring heart rate, and the *hitoe* transmitter 01 for transmitting measured bio-signals to a smartphone **(Fig. 16-1)**.

A feature of the *hitoe* electrode inner shirt comes from the material's high hydro-philicity. During sports and other activities that cause the wearer to sweat, the material absorbs the sweat quickly to continue meas-uring signals in a reliable manner. Field tests are already underway to demonstrate how measurements can be taken using *hitoe* wear in a wide variety of sporting activities, from

Fig.16-1

cycling and athletics to baseball, golf, kendo, badminton, figure skating and many more. Biosignals are also being measured in more extreme activities like motorsports.

In one example of *hitoe* in application, tests were conducted to measure the bio-signals of drivers racing in the Indycar series in the United States. *hitoe* fire-resistance clothing continued to measure the driver's heart rate, ECG, EMG and other biosignals under extreme conditions exceeding 300 km/h.

Analyzing the measured data helped to provide an insight into the physical condi-tions of drivers racing under extreme conditions—an application that is likely to be useful for race management, enhancing driver skills and preventing accidents.

Early indicator of cardiac disease, health management on harsh worksites

hitoe is able to monitor information about the body through the wearer's usual lifestyle easily and in a comfortable, without placing any added stress on the body. The material was initially developed with the aim of playing a role as a medical electrode for the early detection and treatment of illnesses. Performance improvements have continually been made for the *hitoe* electrode, and in August 2016 the *hitoe* medical electrode and *hitoe* lead wire were registered as general medical equipment.

Then in September 2018, Toray Medical Co., Ltd. released the *hitoe* wearable ECG measurement system for medical use that allows Holter ECG to be monitored over an extended period of time **(Fig. 16-2)**. Traditional ECGs based on Holter monitors required

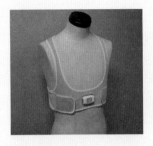

Fig.16-2

the use of adhesives to keep the electrode on the skin, which often resulted in skin problems like irritation and itching. *hitoe* is designed to prevent skin irritation even when worn for long periods of time, to alleviate the stress placed on patients during ECG monitoring. In the future when tests are run at home using *hitoe* Holter ECG monitors, big data containing ECG could be used to detect heart diseases earlier in a way that had not been possible in the past.

More recently, there is growing demand for *hitoe* wear from other areas throughout society, as a means of health management for site workers or to monitor worker safety. For example, workers on construction or similar sites who outdoors for long hours in the summer heat can end up suffering from immense physical stress. They need a system that provides an easy way to monitor their physical condition. There have also been requests for providing a way to support health management and prevent accidents amongst people working alone on night shift as well as long-distance bus and truck drivers.

Toray launched the *hitoe* worker health monitoring service to meet such needs. With this service, the heart rate, acceleration and other data of workers in *hitoe* wear can be measured and analyzed with an app on a smartphone. If conditions are identified that differ to ordinary, an alarm notification function can be used to alert the worker. In emergency situations this allows action to be taken quicker.

As outlined above, the combination of vital data acquired with *hitoe* wear and IoT cloud systems is expected to be applied to a broad range of tools aimed at enhancing the safety and security of humans, in various fields such as medical, sports and safety management.

Case Study 17

Microwell nanobiodevices
Understanding biodata mechanisms and treating incurable diseases

Ever since human gene information was decoded with the completion of the Human Genome Project in 2003, proteins derived from genes and related to a range of life activities have been drawing attention as post-genome biotechnology. Of these proteins, membrane proteins (the proteins adhered to cell membranes) are responsible for the majority of functions of the cell membrane, like information transfer into and out of the cell, and nutrient transport. As such, understanding membrane proteins is becoming a hot topic within the medical field. Amidst these developments, NTT created artificial cells with a microwell structure several micrometers in diameter on a substrate as a model of the most simplified cell. These cells have been shown to work with the functions of membrane proteins.

Membrane proteins with a close relation to human physiological functions

Many membrane proteins have a structure that can pass through cell membranes. There are a number of these proteins that exist: channel proteins that form tiny pores to transport ions and molecules via a concentration gradient; transporters that deform when specific molecules bond to transport substances in and out of the cell; and ion pumps that use energy to transport ions against a concentration gradient.

One example of the function of membrane proteins to focus on is the transmission of information throughout the body by the nervous system. Within the body, a connection called a synapse is formed between neurons to create a network of nerves. Here, the synapse controls the flow of information, with presynapse sending the information and postsynapse receiving it.

Neurons are not connected directly, but rather there is a narrow space (around 20 nm) between them called a synaptic cleft. When an electrical signal reaches the presynapse, a specific chemical is released and the signal is converted to a chemical signal. And when the chemical substance reaches the postsynapse, it bonds with the receiving membrane protein and is turned into an electrical signal again, thus transmitting the information. This flow of information transmission within the body is an essential function for maintaining life, and abnormal information transmission can cause a range of diseases.

In this way, membrane proteins are biological molecules related to many physiological functions, including the onset of diseases, reactions to drugs and immune reaction. There have been many efforts to analyze, or utilize, the functions of membrane proteins using semiconductor devices. Yet most of those efforts did not produce useful results. One challenge with utilizing membrane proteins is that they only trigger their functions while they are within a cell membrane. Simply placing proteins onto a substrate destroys the structure and function of the protein due to a mutual interaction with the semiconductor substrate, which has completely different properties to cell membranes.

Fusing membrane proteins and semiconductor devices

The basic skeleton of a cell membrane consists of a soft lipid bilayer that is self-assembled by the hydrophobic interaction of lipid molecules. Membrane proteins are suspended within the lipid bilayer, and move about slowly within the two-dimensional layer rather than remaining in a single position. We aimed to recreate the same environment as a cell on a semiconductor substrate.

The first step was to fabricate a microwell structure several microns in diameter on the substrate. These wells were sealed by an artificially formed lipid bilayer, with membrane proteins placed inside the linked section. Each of these well structures are approximately the same size as a cell. As the suspended membrane proteins are free to spread within the lipid bilayer sealing the microwells, this is deemed the most simplified model of a cell structure **(Fig. 17-1)**. We have actually proven that these artificial cells operate with the function of membrane proteins.

The illustration below outlines the results when using a channel protein called α-hemolysin. A calcium ion indicator was injected into the well that emits fluorescence in the presence of calcium ions. Some time after adding calcium ion to the outer fluid, fluorescence was observed within the wells. This is thought to be due to calcium ions flowing into the well via the α-hemolysin channel suspended within the lipid bilayer.

Fig.17-1

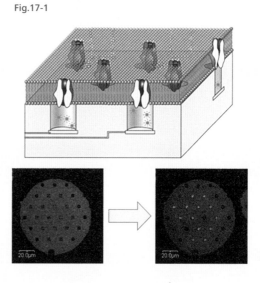

The number of ions flowing into the well was several hundred per second, which makes it possible to perform extremely sensitive measurements. This is because of the small volume of the well on aL scale (attoliter, 10–18 liters), where a tiny inflow of ions can cause a major change in concentration within the well. This level of detection is even smaller than other methods of measuring current (on the level of pA), and applications for ultra-high sensitive measurements is anticipated.

There are still many key problems that need to be addressed in order to develop microwell nanobiodevices other than that outlined above: what type of membrane proteins can be suspended in the lipid bilayers; how does the boundary image appear between the lipid bilayer and substrate; what is the diffusion behavior of membrane proteins or lipid molecules within the cell. To find out, we are performing fundamental research from experimental as well as theoretical perspectives.

Any development of nanobiodevices that fuse membrane proteins and semiconductors is likely to contribute to a broad range of applications in medical, environment and other fields. In other words, creating model cells in a controlled manner within an artificial environment allows us to understand the mechanisms or functions in as simple a way as possible. It also allows information at a molecular protein level to be acquired compared to past methods of finding information on a tissue or cell level. This may help us to better understand the biological information transmission mechanism at the molecular level and find ways of treating incurable diseases.

10 Advanced Materials

Broadening the concept of materials to develop diverse, multifunctional materials

Research and development of advanced material has been underway in many countries around the world, and there are a number of nanomaterials that have already been in practical use as exemplified by carbon nanofibers and titanium oxide nanoparticles. The market size for such materials is forecast to grow into the future, with the market in Asia Pacific region in particular slated to expand at a high rate of 16% per year, much more rapidly than in Europe or the U.S. [1].

Just like many other technologies currently drawing attention, advanced materials are anticipated to be utilized in a wide range of industries. Examples of such advanced materials include: carbon fibers in the manufacturing industry that allow products to be made lighter while retaining strength; self-curable materials in the construction and real-estate fields that may enable automatic repairs of steel and concrete structures; and materials for nanomachines in the medical field that could make it feasible to design and construct *ad arbitrium* nanomachines and to efficiently deliver drugs into the body.

In addition to new materials capable of updating existing products and systems, there is growing need today for materials with new functions suited to harnessing newly developing technologies. Yet there are many challenges remaining to create such materials. In particular, biomaterials are still in the budding stages of development, and any practical applications are likely to take time to develop.

In order to take a major step forward in materials technologies, research into accelerating materials development and creating materials with personalized functions has been drawing attention as described in the followings.

Accelerating materials development by exploiting machine learning techniques

Development of materials has traditionally relied on the experiences of individual researchers. However, this methodology alone is getting more time- and cost-ineffective as synthesis methods and designated material structures become increasingly complex. Now demand for accelerating materials development with the aid of machine learning is evident. In the future, functional materials are expected to be developed in a time-effective manner by exploiting machine learning and data analysis methods capable of putting forward promising candidates for synthesis.

For example, a 2016 article published by Rafael Goemz-Bombarelli et al. at Harvard University outlines how machine learning could be used as virtual screening technology of molecules [2]. In this research, a virtual library of molecules was created using a molecule database, and, with a combination of a machine learning method, the researchers were able to narrow down a range of target molecules that potentially show desired functions. Experiments were then carried out in order of priority. This approach actually allows for time- and cost-effective materials creation.

Applying the technology to the body's drug delivery systems, or regenerative medicine

The second challenge, developing materials with personalized functions, will require development of more complex materials in the future. Until now, even when unique materials were developed, the understanding of microscopic origins of their properties at a molecular level was limited, which often resulted in exploiting only single or one-dimensional functions of a particular material. A deeper understanding of materials at a molecular level will allow us to develop more diverse, multidimensionally-exploitable materials.

One of the most frequently cited papers in recent years in the field of advanced materials was published in 2016 by Matthew J. Webber et al. of Massachusetts Institute of Technology and addresses these challenges [3]. The research focused on the growing demand for biomaterials in the medical field. In addition to the interest in increasing numbers of patients suffering from cancer and wasting disease, there is growing interest in enhancing the QOL of patients of any disease. Molecular target treatment and cellular therapy for reducing stress require advanced materials that have excellent biocompatibility and flexibility and that are capable of changing their desired form and function. A plethora of research is being conducted for this purpose, for instance, on supramolecular biomaterials that can respond to and mimic the body's signal transmission systems, and also deform their shape in a reversible manner. This type of research will bear fruits and culminate in development of the body's drug delivery systems and regenerative medicine in the future.

Developing multifunctional bio-based materials

NTT has also been channeling efforts for researching the above-mentioned subjects as we move ahead with development of new advanced materials. For example, there are efforts for utilizing biomolecules or soft materials to better understand the fundamental principles of information processing deep inside the body. We are aiming to apply those principles to development of highly sensitive biosensors or multifunctional biomaterials, as well as creation of interfaces capable of directly accessing the body.

If new materials can be developed in the field of biomaterials, multifunctional materials could be used naturally in our everyday lives. Examples include implant materials to augment our bodily functions, soft bioelectrode materials capable of constantly sensing information from the body, and sensors that can quickly detect matter like viruses, antibodies or pollen particles. For technology to take on a more natural form, research and development of advanced materials is one of the most appealing topics. We will further accelerate and broaden such R&D efforts toward diverse and multifunctional materials that can be exploited throughout a broader range of fields.

[1] Allied Market Research, Smart Material Market, 2014-2021

[2] Paper published in "Nature Materials" Vol. 15 (Rafael Gomez-Bombarelli et al., "Design of efficient molecular organic light-emitting diodes by a high-throughput virtual screening and experimental approach").

[3] Paper published in "Nature Materials" Vol.15 (Matthew J. Webber et al., "Supramolecular biomaterials").

Case Study 18

Return-to-the-earth battery technology
Addressing the challenge of the IoT era—environment pollution caused by sensors

With the coming of A Trillion Sensors society, a trillion sensors are set revolutionize society, those sensors will be installed everywhere, but will they ever be recovered? With the increasing pace that IoT is becoming commonplace, trillions of devices will be operating around the world. These myriad sensors collect information on a wide range of items and actions to visualize, digitalize and transmit via networks to be analyzed as big data or processed by AI. With this we are able to control and predict items and actions, and achieve a level of efficiency never attained before. All this creates new value that leads to an exciting lifestyle, and which has brought about immense change to our daily lives as well as society as a whole. Conversely, sensors strewn about the place and embedded batteries remain unrecoverable, and there is the risk of them being discarded or left to deteriorate. With this in mind, NTT is developing back-to-earth battery technology.

Using only fertilizer components and bio-derived materials
When sensors are connected by cables, it is possible to trace those cables back to recover any sensors or batteries. Yet wireless sensors mounted to animals, or those installed in places that are difficult to access make it hard to recover the sensors and batteries. Simply leaving them as is can have a significant negative impact on the soil or living matter (Fig. 18-1).

Ordinary batteries such as lithium-ion batteries that are widely used today have a longer operating life and a high level of output due to the increased uptake of electronic devices. They are safety concerns due to their potential flammability, but they also contain many valuable rare metals or toxic substances. These substances contain matter that does not originally belong in the soil, so if batteries become damaged and left in the soil, they may end up cause contamination or affecting lifeforms.

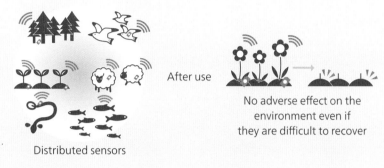

Fig.18-1: Difficult-to-recover sensors and how to solve the issue

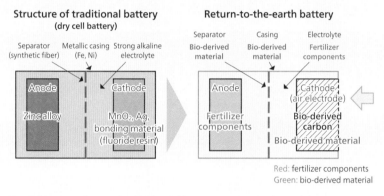

Fig.18-2: Structural materials of traditional battery an Return-to-the-earth battery

To address these concerns, we embarked on research into the Return-to-the-earth battery that uses only fertilizer components and bio-derived materials. The materials used in this battery **(Fig. 18-2)** contain components used in fertilizer, and as such have no adverse effect on the soil even if they are difficult to recover. They are also unlikely to affect plant germination or growth.

Several basic elements like nitrogen, phosphorus and potassium are required as fertilizer components at the correct stages of plant germination and growth. NTT is harnessing the knowledge it has built up of battery materials over the course of R&D into power sources for storage batteries and smartphones used in information communication systems. The battery is designed to induce a chemical reaction from certain combinations of these elements. Research into such batteries began in 2016, and a PoC (Proof of Concept) battery was completed in 2017. This has been registered under the trademark Return-to-the-earth battery.

Carbon electrodes without bonding material

Battery electrodes require a three-dimensional conductive porous structure that oxygen in the air diffuses through. Traditional electrodes are made when a slurry of powdered carbon material is solidified, where fluoride resins or high polymer resins used as the bonding material generate toxic gases if they are burnt. These substances are not usually found in the soil, so they cannot be considered low-polluting materials. To avoid impacting the environment, bonding material must be non-toxic, or electrodes must be made without using bonding material.

We successfully made carbon electrodes free of bonding material, by using bio-derived materials and applying a special pre-treatment process to create a porous structure. When we verified the operation of the battery, we achieved battery performance with a voltage of 1.1 V and discharge capacity of 76 mAh with a measurement current of 1.9 mA/cm2 **(Fig. 18-3)**.

We understand that to make these batteries available for use in various applications, we need to increase their performance to an equivalent level as dry cell batteries that are already commercially available. At the moment we are focusing our research efforts on developing elemental technology to achieve a higher battery voltage,

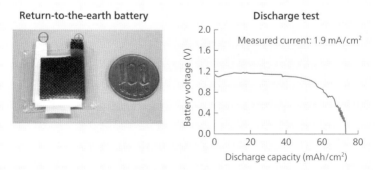

Fig.18-3: Return-to-the-earth battery performance

capacity and layering. We tested the operation of this battery after improvements were made to achieve a low-environmental impact, disposable sensor. The battery was connected to a commercially available BLE (Bluetooth Low Energy) temperature sensor module, and the signal from the sensor module received by a device to check if the battery could operate the sensor module.

Development of sensor non-toxic to soil

A plant toxicity test based on assessment methods for fertilizers was used to check whether the battery affected plant life. Used batteries were crushed into powder and mixed with soil, into which mustard spinach seedlings were planted and tested for germination state. The plant toxicity test is a method used to assess the toxicity of components in fertilizer or soil by observing the growth conditions of plants. The test results indicated that the Retun-to-the-earth battery® differed to traditional batteries (battery A) in that it had no impact on the growth of plants. In other words, the growth rate was observed to be the same as plants grown in soil without any battery material mixed in, demonstrating the success of the Return-to-the-earth concept **(Fig. 18-4)**

Future challenges include increasing the performance of the battery as well as applying NTT's forte with LSI circuit design technology to develop back-to-earth sensors and circuits. Applying this to distributed sensor services using disposable sensors will help to contribute to and achieve an IoT society that offers an affluent and convenient lifestyle well-balanced with nature.

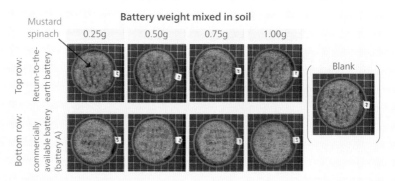

Fig.18-4: Impact on plant growth

Case Study 19

Semiconductor hetero-nanowire
Opening up the potential of new semiconductor devices

Current semiconductor devices used extensively in computers and communications equipment. One of the fundamental structures used in such devices is the hetero-structure, where different types of semiconductors are bonded together in thin layers. In the past, the types of semiconductors that could be formed in a hetero-structure were limited, and high-precision top-down fabrication technology was required as devices became smaller with higher density integration. If there were to be no limit to the types of semiconductors that could be connected together, diverse range of functions could be developed that were not previously available. And if compact devices could be fabricated without needing high-precision fabrication technology, high-density integrated devices could be developed at low cost and without the risk of damage from the top-down process. In light of this, NTT is employing a structure dubbed semiconductor nanowires to redefine common knowledge and develop technology capable of fabricating new nanoscale hetero-structures and devices.

Semiconductor nanowire of any hetero-structure materials
Semiconductor nanowires are semiconductors with a strand-like structure around 1/100th the thickness of a human hair (several tens to hundreds of nm). Semiconductor crystals have a specific grid length (lattice constant) connecting the atoms depending on the type of material. Ordinary semiconductor hetero-structures require matters with a similar grid length to be connected together, which limits the combination of semiconductor materials.

Hetero-structures using semiconductor nanowires are extremely thin in diameter, so even with slight differences in grid length, different materials are happy to stay bonded together. If these nanowires can be used, the range of combinations of hetero-structure materials will increase significantly.

Semiconductor nanowire structures also do not need to be made with high-pre-

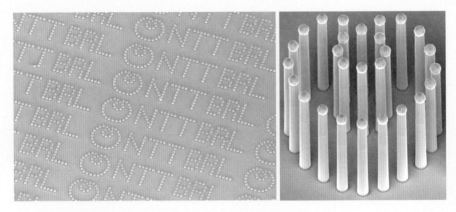

Fig.19-1: Electron microscope image of arranged nanowires

cision fabrication technology, and instead can be formed perpendicularly onto the semiconductor substrate directly using a crystal growth technique. This means there is no damage from top-down fabrication process, and large quantities of nanowires can be made in batches.

To achieve this, particles of metals like gold are arranged on ordinary semiconductor substrates as the seed for inducing crystal growth for the nanowires. Yet gold is considered an impurity for most semiconductors, and there are concerns that it may cause deterioration in device performance. To overcome this, we used particles comprising one of the elements of the actual semiconductor material as nanowire seed, and also developed a method for controlling the position of those seeds at will.

Applications anticipated for photonic integrated circuits, flexible displays and quantum research

One example of the fruits of research into semiconductor nanowire devices using NTT's proprietary technology is the world's first successful operation of a tiny laser emitting light in the optical communications wavelength band from a single nanowire. In this case, indium arsenide (InAs) and indium phosphide (InP) were used as hetero-structure materials. The depth of the InAs light emitting layer was changed precisely to demonstrate that the laser oscillating wavelength could be controlled at will between 1.3 and 1.6 micrometers used in the optical communications wavelength band.

This InAs and InP hetero-structure combination was simply not possible in the past, making this a groundbreaking result. Additionally, there are no concerns regarding degradation of device performance due to impurities, as no gold particles are used.

Applications of this technology are anticipated to lead to development of various nanowire devices. One of these is the possibility of creating nanowire devices on different materials, as nanowires mean any hetero-structure combination can be used as well as forming any type of substrate.

Anticipated examples include directly mounting nanowire light sources onto photonic integrated circuits, or applications like flexible displays where nanowires can be mounted on substrates that can be bent anywhere. And due to their small size, the electrons within a semiconductor nanowire exhibit behavior based on quantum mechanics. These properties may be used in the future to develop non-canonical light sources (single-photon [1] and quantum entangled photon-pair sources [2]), single electron transistors, and even for utilized for Majorana fermion observation [3].

We have already made light-emitting diodes and laser structures using semiconductor nanowire structures, and observed that these function as devices. These are completely new types of semiconductor devices, and may result in breakthrough applications for transmitting and receiving devices used in telecommunications technology, as well as information processing devices like computers and mobile phones.

[1] One type of elementary particle, and in quantum field theory is the name given to a particle of light (electromagnetic wave).

[2] A state where a correlation exists between two photons that cannot be explained by classic physics. Quantum entanglement is the physical phenomenon where the quantum state of one particle cannot be separated from the other. Measuring one particle results in a specific unique property where the state of the other spatially separated particle is also defined. This property is anticipated to be applied to applications like quantum information communications and quantum computing.

[3] An elemental, neutral fermion particle that has the unique property where the particle and its own antiparticle are the same. They were hypothesized to exist in 1937 by Italian physicist Ettore Majorana. In recent years, the existence of quasiparticles that behave the same way in solids has been reported, and these are expected to be the focus of major research as they are likely to contribute significantly to advances in quantum computing.

11 Additive Manufacturing
Growing anticipation for 4D printing and bio-printing

The realm of 3D printing is drawing increasing attention. Yet while aware-ness of additive manufacturing has grown in recent years, the technology is still said to be in its early stages. The term additive method was originally used for printed circuit boards in the manufacturing method where circuit patterns were added later to insulated substrates. This concept has evolved into a three-di-mensional technology, and contributed to the major strides made to high-density and miniaturized semiconductors. Three-dimensional technology subsequently led to additive manufacturing developed by General Electric (GE) stemming from its resin-based 3D printers.

In a narrower sense, additive manufacturing is well-known in Japan as 3D printers, however the technology is budding from the 3D printer sector to span a wide range of fields. The market is growing at a steady pace, and is predicted to reach a scale of $21.5 billion by 2025, mainly led by initiatives in North America and Europe [1].

An emerging technology that goes beyond three dimensions is 4D printing, which incorporates information on changes in time and condition, and potential applicable fields also continue to grow. If 4D printing evolves to become a reality, one practical application is the development of self-healing materials which could revolutionize materials used in the construction industry. In the medical field, prosthetic arms and legs could be made, offering superior adaptability and advanced personalization. And the automobile industry could benefit from flexi-ble component manufacturing potentially leading to faster development of pro-totypes.

Yet of course there many challenges that remain before additive manufac-turing becomes commonplace. The technology is being applied to a broader scope of construction and materials, from manufacturing simple to complex structures, or from using single to multiple materials. Cost reductions will also be necessary so that the technology is accessible to more people. There are three main areas that we are currently focus on to address the issues related to additive manufacturing: material diversification; faster printing; and higher precision printing.

Achieving material diversification and faster printing

The first challenge, material diversification, refers to expanding the variation of materials that can be used. While people are likely to think of materials like resins or metals when it comes to 3D printing, today only few materials can be used with additive manufacturing, and these restrictions have prevented greater adoption of the technology. There are high expectations for expanding the material range to include ceramics and alloys. A greater range of materials will make it possible to fabricate human teeth and bones, and there are even research efforts aimed at making artificial organs with 3D printing.

For example, a paper published by Swee Leong Sing et al of Nanyang Technological University focused on using titanium tantalum composites for additive manufacturing [2]. Titanium tantalum composites are known to be very bio-compatible with human tissue, and advances in this research could lead to practical applications in the medical field (such as bone screws, implants, and devices that need to be embedded in the human body).

Further to this, research conducted by Zak C. Eckel et al at HRL Laboratories, LCC that is owned jointly by Boeing and GM studied the potential of using superior heat-resistant ceramic materials for additive manufacturing. Advances in this research is anticipated to lead to applications in the aerospace industry, like drive components for aircraft [3].

The next challenge, faster printing, addresses the fact that current means of manufacturing takes time as the most common configuration used for 3D printing consists of output from a single head. Research institutions around the world are channeling efforts to develop multiple-head printers, or combining multiple technologies with two-stage hardening as a way of overcoming this issue. Two-stage hardening in particular is a field showing promise. For instance, a 2018 paper published by Xiao Kuang et al of the Georgia Institute of Technology demonstrates how a large reduction in printing time can be achieved with a combination of two material hardening methods using light and heat [4].

Growing anticipation for bio-printing

The third area, higher precision printing, is related to printing precision and quality. With current technology, achieving reliable layering continues to be a challenge, and attention is turning to methods of avoiding fluctuations in the thickness and size of printed layers. Without increases in precision, it goes without saying that it will be difficult to achieve the true advantages of 3D printing or apply them to advanced fields. Many research institutions are focusing on overcoming this issue. In 2018, Andrey Vyatskikh et al of the California Institute of Technology published a paper in "Nature Communications" where they used light exposure technology as a new approach to additive manufacturing [5]. The research teams used this method to successfully create a nanolattice structure on a 100-nanometer level.

As progress is being made in this field of research, NTT is also keeping up the pace by conducting research in the field of 4D printing and bio-printing with a focus on personalized bio-devices.

NTT is also researching and developing hetero-material integration technology for bonding various materials onto silicon—technology that will be required for integrated circuits that converge light and electrons. Photonics-electronics convergence devices require different materials for photons and electrons, and those differing materials need to be bonded on silicon to operate properly. This field is exactly where the concept of additive manufacturing is required—a completely different approach to ordinary 4D printing or bio-printing will be required. It is here that we want to apply our particular expertise.

As additive printing continues to make strides, it will eventually be possible to make devices suited to each and every individual person. The potential of personalized manufacturing will also come to light in a wide range of other fields. Advances in additive printing are expected to make manufacturing even smarter with each step.

[1] Frost & Sullivan, Global Additive Manufacturing Market Forecast to 2025

[2] Paper published in "Journal of Alloys and Compounds" Vol. 660 (Swee Leong Sing et al., "Selective laser melting of titanium alloy with 50 wt% tantalum: Microstructure and mechanical properties"). Titanium tantalum composites are anticipated to be used for applications like bone screws and implants.

[3] Paper published in "Science" Vol. 351 (Zak C. Eckel et al., "Additive manufacturing of polymer-derived ceramics"). Machining ceramic materials had been difficult in the past, however additive manufacturing may help overcome this issue.

[4] Paper published in "Macromolecular Rapid Communications" Vol. 39 (Xiao Kuang et al., "High-Speed 3D Printing of High-Performance Thermosetting Polymers via Two-Stage Curing").

[5] Paper published in "Nature Communications" Vol. 9 (Andrey Vyatskikh et al., "Additive manufacturing of 3D nano-architected metals")

Case Study 20

Self-assembly of biocompatible 3D structures
Creating the desired shapes and functions

In the fields of drug discovery, regenerative medicine and cell transplantation therapy, there is currently great demand for assembling cells into three-dimensional (3D) structures that are close to those of biological tissues. Although cell printing methods and cell cultures in petri dishes were conventionally used to create cell aggregates, their poor precision of shapes and low biocompatibility had presented challenges. To address these issues, NTT developed a new technique to transform polymeric films with excellent flexibility and biocompatibility to any desired 3D shape. This technique made it possible to provide biocompatible molds for encapsulating and cultivating cells, and reconstruct artificial and tissue-like 3D structures.

3D transformation of multi-layered film

This technology uses two or more soft materials that are layered in thicknesses of several hundred nanometers. These layered films are spontaneously transformed (self-folding) into desired 3D structures. The multi-layered film undergoes self-folding distortion where it spontaneously deforms depending on the difference in the stiffness of the film material or film thicknesses. So far, it had been technically difficult to stack different types of soft materials and form strongly bonded multi-layers. In this study, we successfully made a multi-layered and tightly bonded thin film by selecting the most suitable polymer including hydrogel and optimizing the film geometry.

Fig. 20-1 shows the self-folding process of the films composed of silk cocoon-derived hydrogel, called silk fibroin. The polymer film called poly(p-xylylene) (parylene) was laminated on the surface of silk fibroin film with chemical vapor deposition, and micro-patterned with two-dimensional (2D) shapes using a photolithography technique.

Releasing these micro-patterned bilayer films from a substrate triggers a

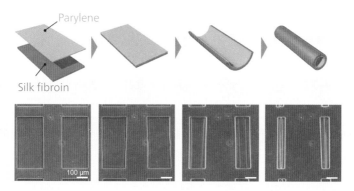

Parylene

Silk fibroin

100 μm

Fig.20-1

self-folding process in which the rectangular films were peeled from the edges, bent in the direction of the shorter axis without any additional force applied, eventually forming a cylindrical structure. In addition to rectangles, films with irradiation patterns were self-folded to create spherical structures, and ones with rectangle patterns connected with hinges were self-folded to create doll-like structures. In this way, the final self-folded 3D structures were found to be totally dependent on their corresponding 2D micro-patterns.

Encapsulation of cardiomyocytes

All the polymers constituting the film used in this research show no toxicity to cells, and thus can be used as an interface with biological samples like cells or biological tissue. When the cells were suspended in advance on the surface of multi-layer films, they were able to be encapsulated inside 3D shapes as the film undergoes self-folding, without causing any damage to the cells **(Fig. 20-2)**.

Armed with this knowledge, we succeeded in incubating cells inside the 3D structures, and forming cell aggregates with unique functions that are similar to those of biological tissue. Growing cardiomyocytes within a cylindrical structure over a period of around

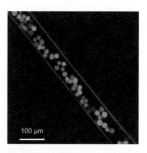

100 μm

Fig.20-2

one week results in forming cell aggregation lengthwise along the cylinder, and creating a structure functioning as a single tiny myocardial fiber. These reconstructed myocardial fibers were observed to have the same unique functions as cardiac tissue, like cells beating in sync with others, as well as cell-cell interactions such as calcium oscillations at the same period as their beating.

This method helped us to make various types and shapes of 3D templates with a high degree of design flexibility and curvature radii, by controlling the elastic moduli and film thicknesses. In addition to cylindrical shapes and cardiomyocytes, different types of 2D micro-patterns and cells lead to the reconstitution of biological tissue-like structure with various 3D shapes and specific cell types.

Applications in regenerative medicine and medical implants

The self-folding technology presented here is not limited to silk fibroin, and can be adapted to a wide range of functional materials. For example, graphene, which is a single layer of carbon atoms formed in a 2D hexagonal format, could also trigger self-folding of parylene films when it is transferred onto the parylene surface **(Fig. 20-3)**.

We applied this phenomenon to encapsulation and cutlivation of neurons within a 3D cylinder of graphene-parylene bilayer. The encapsulated neurons gradually connected each other and formed cell aggregates during the long-term incubation process. We also observed that they extend their neurites both inside and outside of the cylindrical structure in order to exhibit their intrinsic cell-cell interactions **(Fig. 20-4)**.

The reconstituted neural tissue formed here was observed to respond to the ap-

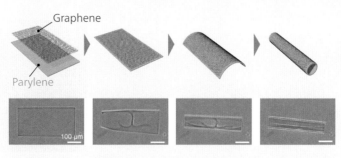

Fig.20-3

plied external stimulus simultaneously, which indicated that cellular information is being correctly transmitted between the adjacent cells within this tiny tissue. Graphene, meanwhile, is made up of only carbon atoms, but its features include optical transparency, mechanical rigidity, and biocompatibility as well as electrical conductivity. Therefore, this self-folding technology to form a 3D structure of graphene is applicable for developing a tiny biocompatible 3D electrode to investigate the electrical properties of the cells and tissues.

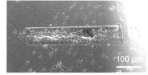
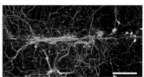

Fig.20-4

In this research, we focused on the theory behind self-folding of the films, and discovered a method for assembling 3D templates that are flexible and compatible with biological samples like cells and biological tissues. This method allows many types of tiny artificial tissues to be reconstructed, and may lead to a regenerative medicine platform that is indispensable for drug discovery and cell implantation therapy. Other possibilities include the tools for analyzing the behavior of single cells in confined spaces. Furthermore, from the perspectives of material science, additional functions such as electrical conductivity and response to magnetic fields will broaden the variety of developed biological interfaces including implantable bio-electrodes or actuators.

Researchers(Part2)

CS1. Speech recognition and casual conversation technologies
Koh Mitsuda, Mana Ihori, Yuki Kitagishi and Mizuki Nagano, NTT Media Intelligence Laboratories

CS2. Angle-free Object Information Retrieval Technology
Jun Shimamura, NTT Media Intelligence Laboratories

CS3. Spatio-temporal Multidimensional Collective Data Analysis Technology
Akira Nakayama, NTT Service Evolution Laboratories

CS4. *Kirari!*, ultra-realistic communications
Shinji Fukatsu, Hideaki Takada, Kimitaka Tsutsumi and Yoshihide Tonomura, NTT Service Evolution Laboratories

CS5. *Point of Atmosphere*
Kenichi Minami, NTT Service Evolution Laboratories

CS6. *Deformation Lamps* and *Hidden Stereo*
Takahiro Kawabe and Taiki Fukiage, NTT Communication Science Laboratories

CS7. *Piper*, botnet detection infrastructure employing large-scale flow analysis
Akira Morikawa and Kazunori Kamiya, NTT Secure Platform Laboratories

CS8. *LRR*, platform for collaborative defense against cyber defense
Eitaro Shioji, Kazufumi Aoki and Kunio Hato, NTT Secure Platform Laboratories

CS9. AI optical interconnect technology
Takeshi Sakamoto, NTT Device Technology Laboratories

CS10. Optical al Fiber for Space Division Multiplexing
Kazuhide Nakajima and Takashi Matsui, NTT Access Network Service Systems Laboratories

CS11. Integrated optical front-end device technology
Hideyuki Nosaka, NTT Device Technology Laboratories
Shunichi Soma, NTT Device Innovation Center
Yutaka Miyamoto, NTT Network Innovation Laboratories

CS12. Artificial photosynthesis technology
Takeshi Komatsu, NTT Device Technology Laboratories

CS13. *LASOLV*
Hiroki Takesue, NTT Basic Research Laboratories

CS14. Proof-of-principle experiment for quantum
       repeaters using all-photonic quantum repeater scheme
Koji Azuma, NTT Basic Research Laboratories

CS15. Nanophotonic device technology
Kengo Nozaki, Akihiko Shinya and Masaya Notomi, NTT Basic Research Laboratories

CS16. *hitoe*® performance material for collecting biosignals
Hiroshi Nakashima and Masumi Yamaguchi, NTT Basic Research Laboratories

CS17. Microwell nanobiodevices
Yoshiaki Kashimura, NTT Basic Research Laboratories

CS18. Return-to-the-earth battery technology
Takeshi Komatsu, NTT Device Technology Laboratories

CS19. Semiconductor hetero-nanowire
Guoqiang Zhang and Takehiko Tawara, NTT Basic Research Laboratories

CS20. Self-Assembly of biocompatible 3D structures
Tetsuhiko Teshima, NTT Basic Research Laboratories

Supervising Editor & Authors

Supervising Editor
Jun Sawada

President, Chief Executive Officer, and Member of the Board, NTT. Joined Nippon Telegraph and Telephone Public Corporation in 1978. After working in areas like technology development, service development, corporate sales, and corporate planning, became Senior Executive Vice President and Member of the Board of NTT in 2014, and assumed his current position in June 2018.

Authors
Motoyuki Ii

Senior Executive Vice President and Member of the Board, NTT. Joined Nippon Telegraph and Telephone Public Corporation in 1983. Became Manager of the Niigata Branch of NTT East in July 2007, Senior Vice President and concurrent Executive Manager of the Plant and Planning Departments of the Network Business Headquarters of NTT East in June 2011, Senior Executive Vice President and Senior Executive Manager of the Corporate Sales Promotion Headquarters of NTT East in June 2016, and assumed his current position in June 2018.

Katsuhiko Kawazoe

Senior Vice President, Head of Research and Development Planning, Member of the Board, NTT. Joined NTT in 1987. Became the Chief Producer of content distribution services and Vice President of Research and Development Planning in 2008, Head of the Service Evolution Laboratories in 2014, Head of the Service Innovation Laboratory Group in 2016, and assumed his current position in June 2018.